THE PHARON INVADE

On the screen Stephanie could see the Pharon warriors shambling up the street in their ornate, golden armor.

The walking dead. The Rotten.

Their armor was beautiful, covered with symbols and hieroglyphics, and with high ornate collars behind their heads. But their faces were nothing more than wrappings and gray skin, long since dead, as if frostbite had taken them. The warriors were in pretty good condition for dead beings. Only one or two had complete chunks of them missing or rotted away.

Suddenly on the display the missile streaked in and smashed straight into the middle of the group, exploding in a surrealist sort of slow motion. On the screen it was clear that at least four of the aliens had been blown completely apart by the missile explosion.

At that moment three of the other warriors took aim upward with their weapons.

"They aren't going to retreat from this," Stanton said.

VOR: THE MAELSTROM

Novels available from Warner Aspect

Vor: Into the Maelstrom
by Loren L. Coleman

Vor: The Playback War
by Lisa Smedman

ISLAND OF POWER

DEAN WESLEY SMITH

ASPECT®

WARNER BOOKS

A Time Warner Company

WARNER BOOKS EDITION

Cover design by Don Puckey
Cover illustration by Donato
Cover logo design by Jim Nelson

Warner Books, Inc.
1271 Avenue of the Americas
New York, NY 10020

Visit our Web site at
www.twbookmark.com

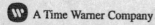 A Time Warner Company

Printed in the United States of America

First Printing: April 2000

10 9 8 7 6 5 4 3 2 1

For Loren Coleman

Thanks for all the friendship

ISLAND OF POWER

Prologue

**Time: 12:14 Universal Time
17 minutes before Arrival**

The Pharon high priest stood in the control room of his spacecraft, the jewel-of-power solar collector over his head almost touching the ceiling as he stood watching the viewscreen. Displayed before him against the blackness of space was what looked like a small asteroid, no more than six kilometers long and half that thick. The light of the Maw, distant and faint, traced only the rock's outline against the few stars. This was an almost empty area of the Maelstrom, near the outer edge. There was very little here to interest his people except this rock.

A simple-looking asteroid tumbling in the darkness.

But the high priest knew that it was much more. The rock contained vast power. A power he would someday control for his people.

The fluids of life coursed through his body, pumped into him by the pack on his back. The wrappings on his arms were new and uncomfortably tight, yet he knew he must keep them that way. His vest was also new, with ornate glyphs showing his rank and accomplishments. His mailed skirt hung down behind him, protective rear armor. Soon, when he controlled the power of the rock, he would have even more beautifully carved glyphs to be proud of on his vest.

His craft hung in space to one side of the tumbling asteroid, the ship's sleek, aerodynamic shape giving the impression of movement even when it was hovering. The nose of the ship was pointed slightly downward, looking like a bird's head from a distance. The gold and chrome of its skin reflected the distant light of the Maw at the center of the Maelstrom like a faint, internal glow.

At least twenty other Pharons, mostly foot soldiers, also accompanied the high priest, as well as the two pilots sitting to his right. But he ignored them all for the moment, spellbound by the slowly tumbling rock showing on the screen.

Suddenly a blue light seemed to encircle the rock, illuminating structures on the surface. One entire side of the asteroid was flat and covered in tall buildings never intended to exist in the vacuum of space.

A city.

An alien city could now be clearly seen. He knew the shape of it from memory, the pattern of its streets, the height of its buildings. He had followed this city for longer than he dared remember, always hoping that one day he would get the chance to learn its secrets. But as long as it remained in the coldness of space, the city's secrets remained just that.

Secrets.

"Move back and stand ready," he ordered the pilot, who obeyed, maneuvering away from the blue glow surrounding the city-covered asteroid.

The glow increased, expanding, with no clear power source.

Then the glow contracted.

Then expanded again, the entire time never more than a kilometer off the surface of the rock.

The Pharon priest marveled at the power of it. He must learn what controlled such power, then tame it.

The city became clear, brightly lit as if spotlights were covering its every centimeter. Towering structures seemed to reach for the ball of blue energy surrounding the asteroid, never quite touching it.

Then the blue glow began to swirl with intense force, expanding, contracting in the void of space.

"It is about to shift," the high priest said to the crewmen beside him.

The blue of the energy surrounding the city-covered rock seemed to flash brightly as it reached its most intense point.

In an instant the rock and the alien city winked out of sight, leaving only the faint shadow of a blue ball that also quickly vanished.

The Pharon priest stared for a moment at the

empty viewscreen, then turned to his second. "Where did the city jump to?"

"A new planet's surface," his second said. "Very close to our current position."

For the first time in hundreds of jumps, the city had landed on a planet. And it was close enough to reach in time. The high priest's voice concealed his excitement.

"Take us there," he said. "Quickly."

In the blackness of space the sleek shape of the Pharon craft turned and shot off, tracking the alien city once again to yet another new destination in the Maelstrom.

1

D r. Hank Downer shivered. The wind coming
off the Pacific Ocean cut through his light
jacket as if it wasn't there, the salt air sting-
ing his neck and face. He pulled the zipper up a little
closer to his chin, hoping to block even the smallest
breath of wind. It did no good. He would never get
used to the dampness of the ocean air and the biting
cold of the Oregon nights. The weather there never
seemed to change. Cold rain and cold wind. Always
the same.

At least at the moment there was no rain. But the
wind felt like it was peeling off layers of skin. The

noise of the wind against his ears threatened to drown out his thoughts, but it was not enough to overpower the sound of the pounding of the surf.

He and Stephanie Peters stood on a beach that stretched for kilometers of blackness in either direction. To the south a few lights shone from the small coastal town of Dustin Cove. Once a thriving coastal fishing community, the town was now populated mostly by research scientists like him and Stephanie and the facility's military contingent and their families. They all worked at the nearby Oregon Research Facility. The Union government had long since bought out the original residents, more than likely for security reasons.

In the other direction was nothing but sand and rock bluffs. This was a desolate stretch of land, isolated from the population centers of Oregon by mountains. It must have been the isolation that had made the government decide to place the research center out there. But the government didn't have to live there.

"Isn't it a beautiful night?" Stephanie asked, putting her arm through his and pulling herself up close to him. "Smell that wonderful ocean. After all day cooped up underground, this is great."

It was a little after one in the morning, and they both had to be back in the lab at eight. If it were up to him, he'd be asleep right now in a warm bed. But she had wanted to take a stroll on the beach, to "clean some of the cobwebs out of my mind," she had said. He hadn't wanted her to go alone. Besides, it was a rare chance to spend a little more time together.

So there they were.

Stephanie stood facing directly into the wind and took a long, deep breath. Then she pointed at the black sky. "There are actually a few stars out."

"Very few," he said. The wind seemed to yank the words from his mouth and float them up the sand dune and into the pine forest behind them. The paucity of stars in the clear night sky was just another of many reminders that Earth wasn't where it used to be. Six years ago the planet had been ripped from its place among the stars and sucked into the anomalous space everyone now called the Maelstrom, a place where the laws of physics seemed to take every other day as a holiday.

What Stephanie was calling stars he knew were mostly just planets that had been pulled in along with Earth, close enough to reflect the bright light coming from the Maw.

She knew that, too, but Stephanie still liked to pretend they were stars and that the night sky was beautiful. Not Hank. He could never forget the thousands of stars that once filled the heavens and the fun of picking out the constellations or watching Mars and Jupiter move through them. For him, nothing could compare with that. These few points of faint light, scattered like dots on a black page, would never be enough to be beautiful to him.

"Oh, come on," Stephanie said, giving his arm a squeeze. "Let's just enjoy a little of the night and not think about anything for a few minutes."

She let go of his arm and headed down the beach, her small flashlight making sure of the path ahead of her.

He followed, staying close in the darkness, not turning on his larger light, but using hers for his guide as well. His feet sank with every step into the soft sand. He could feel the grains flowing in over the top of his boots and grinding through his socks. By the time he and Stephanie had gone the hundred meters south to a small alcove of sand, rocks, and driftwood, he was limping. Both boots felt at least a couple of kilos heavier.

Stephanie dropped down into the sand near the bottom of a steep dune facing out over the dark ocean. Then she clicked off her flashlight. Hank stopped to sit on a log a few steps away and started untying his boots. At least here the wind wasn't numbing his ears and face. And he could almost hear himself think, even with the pounding surf just down the beach.

His eyes had grown accustomed enough to the faint light to see that she was staring out over the ocean, her chin resting on her pulled-up knees.

"I love it out here," she said. "It makes me forget all my cares."

Hank didn't answer. Again, it didn't actually work that way for him. Most of the time he just kept right on worrying no matter where he was.

He left his boots untied and moved over to sit down beside her. She instantly leaned against him, resting her head against his chest while continuing to look out over the ocean.

He put his arm around her. Though they'd been together for six months, they were rarely alone together. In their work in the underground research bunker, they were surrounded by hundreds of other scientists also working there.

"That's nice," she said, snuggling against him.

He had to agree. It was. Very nice, actually.

They sat there, listening to the wind and the roaring surf for what seemed like a long time, saying nothing, not even moving. Sometimes moments were just meant to be left that way, Hank thought, and this was one he would always remember.

Suddenly Stephanie sat upright, pulling out of his grasp and pointing out over the ocean.

He saw it, too.

The surface of the ocean was now visible before them, as if lit from above. He could see the pounding waves, the black rocks, the wind-whipped whitecaps.

"What's causing that?" he asked.

"Not a clue," she said.

Around them it kept getting brighter and brighter, the light almost white as it turned the night brighter than a well-lit lab.

Suddenly the sky over the ocean seemed to crackle, and everything around them had a bluish tint. The waves, the sand, the rocks.

Everything.

Then the sky over the ocean exploded.

Lightning struck the top of a wave a short distance offshore. The bolt came out of nowhere. There were no clouds, nothing but the blackness of the night sky and the intense white light seeming to come from everywhere at once.

"Amazing," Stephanie whispered.

Hank, too, was in awe. He'd never seen anything like what was happening around them.

Then more lightning strikes.

And more.

And more. Faster they came, until it seemed the bolts were flashing down everywhere, as far up and down the beach, as far out over the ocean, as he could see.

And they were coming at them.

He could now both feel and sense the electrical charge in the air around them.

His clothes seemed to grow stiff, crackling with little sparks.

His hair floated away from his head.

Electricity, massive amounts of it, was flooding over the ocean and beach.

He pulled Stephanie back, and they scrambled on hands and knees, staying close to the ground, until they were lying behind the dune. There were no rocks or logs around them. Just sand. A dune didn't make a good place to hide, but it was the best they could do without trying to stand and run. And Hank had no doubt that if they did that, they'd soon be dead.

Thunder shook the ground as more and more lightning struck up and down the beach and over the waves. Hank and Stephanie wiggled their way down into the sand, trying not to be even a slight bump above the surface.

One strike hit the log right where Hank had been sitting a few moments before and covered them in sand and splintered wood.

His ears rang from the unbelievably loud concussion.

"That was too close!" he shouted.

Stephanie only nodded as the deafening thunder filled the air around them.

Vast plumes of water shot into white light with each lightning strike on the ocean. And what had been night had turned into the brightest day. The air itself seemed to glow.

"What's happening?" Stephanie shouted, her voice barely audible between the intense claps of thunder.

He had no idea what was happening. Plenty of strange things had been occurring all over the planet since that awful moment when the Earth was drawn into the Maelstrom, but Hank had never been this close to any of them. Nothing like this.

"I don't know!" he shouted. "It must be something from the Maelstrom."

It was the only thing he could think of. Nothing like this ever happened naturally on Earth. What else could it be but the Maelstrom?

During the first days of the Change, the Earth had suffered myriad storms, cataclysms, and general disasters. A large chunk of United Africa had even vanished, but things had eventually settled into a kind of normality. Of course, that didn't mean nothing else was happening. It just meant it wasn't getting reported to the general populace. Even he and Stephanie, working at a top-secret Union research facility, knew only what related to their own projects.

Lightning and thunder pounded the beach around them.

Another strike of lightning hit the hill above them, covering them with even more sand. He felt like they were in the middle of a giant battle of the gods. Everything was shaking.

Sand rattled down the slope above them.

Ocean water sprayed them as the wind drove it inland.

Lightning continued to strike all around, and thunder rocked the ground.

They didn't dare move.

Hank felt like he didn't dare even breathe.

He held Stephanie, and she held him back as the world around them went totally and completely crazy.

Out on the horizon, through the bright flashes, Hank thought he could see a dark shape floating downward in the bright light.

"Over the ocean!" he shouted into Stephanie's ear. "Look."

Stephanie nodded. She was watching it, too.

The black shape was like a hole in the white light.

Slowly, like it was drifting down on a slight breeze, the hole in the light settled into the ocean, just short of the horizon.

Then, almost as quickly as the storm had begun, the lightning slowed down. A few more strikes on the waves, and it stopped completely. The rumbling thunder, too.

Hank felt like a weight had been lifted from his chest. His entire body was vibrating.

With one last rumbling crack, the air returned to

being a simple ocean breeze, now thick with the smell of ozone mixed with brine and fish.

Then the white light faded, as if sucked out over the ocean to a point where the black shape had been.

The point of light remained there for a few long seconds, like a very short sunset of pure white energy.

Then it was dark again.

Pitch-dark.

Hank couldn't see a thing.

Not even Stephanie just centimeters in front of his face. His eyes had adjusted to the bright light, and it was going to take some time for them to adjust back to the dark, starless night.

They both lay still for a time, as if expecting the wild storm to come back. But there was only darkness and the normal sounds of the ocean waves on the beach.

Hank sat up out of the sand and fumbled in his jacket pocket. He managed to pull out his flashlight at the same moment Stephanie turned hers on. The beams seemed almost pitiful in the darkness compared to the white light of a few moments before. They both shone their flashlights out over the ocean, but could see nothing. The beams were swallowed by the blackness.

The log where Hank had sat down to empty the sand from his boots was a smoking hunk of splintered wood. He didn't want to think about what he'd look like if he'd still been sitting there.

"Amazing," was all he could think to say as they stood up. He brushed the sand out of his hair and off his pants.

Stephanie just kept staring out over the ocean.

Quickly he bent and tied his boots, then said, "Quite a light show, wasn't it?"

"We've got to get off this beach," she said suddenly, her face still turned toward the ocean. "If that black thing we saw was something hitting the ocean out there, we've got a tsunami coming."

He knew instantly that she was right. Even before the Earth had been sucked into the Maelstrom, tsunamis regularly struck the West Coast of North America. Most were small, generated by earthquakes in Japan or Alaska. But some had been large, killing dozens of people. This beach was the last place anyone wanted to be when even a small tsunami arrived.

They'd been briefed on earthquakes and tsunamis shortly after arriving at the facility because a major fault line ran just a few kilometers off the coast. The memory of the hour-long session on tsunami survival came rushing back to him. At the time he thought it almost a waste of time. The subject had been laughably simple. "If you think a tsunami is coming, get to high ground."

Then the rest of the hour had been filled with the speed of waves and destructive power of the waves. The only actual new fact he'd learned was that a tsunami wasn't just one big wave, but a series of them. How fast the first wave would reach the shore would completely depend on how far out the earthquake was. Or how far out something hit the ocean. Tonight it looked as if something just might have done just that.

If it was only two kilometers out, they had five

minutes to get to high ground. If the black shape hit ten kilometers out, they had a little longer. He figured thirty seconds to a minute had already gone by since they'd seen the shadow come down.

"Run for it," he said. "Trade flashlights. I'll be right behind you."

He handed her his larger flashlight and took her smaller one, then glanced at his watch before taking off along the beach in front of the dunes.

Running the hundred meters back up the beach in soft sand seemed to take forever. Hank kept glancing out over the ocean, his eyes watering in the wind, looking to see if the waves were being sucked back, the first indicator of a large tsunami wave coming in. But the moonless night was too dark to see anything.

They reached the path up the sandbank, and Stephanie scrambled upward without stopping. She was almost ten years younger and in much better shape than he was. He forced himself to stop, take two deep breaths, then start up behind her.

The research facility was buried underground in the side of the mountain inland from the beach. The huge blast doors there would protect them from any large waves if they could make it there in time.

At the top of the dune she waited for him, working to catch her breath.

"You all right?" she asked, panting as he scrambled up beside her.

He could only nod because he didn't have enough wind to even think about talking. The sand again filled his boots and socks, but at the moment there wasn't time to empty them.

Stephanie shined the light on her watch. "It's been almost two and a half minutes," she said. "Let's go."

"Right behind you," he managed to say between choking breaths of cold air.

The ground became harder, packed down and much easier to run on. Stephanie set a solid pace, one that allowed her to keep her flashlight on the path ahead. He stayed right behind her, even with his sand-filled boots. Hank was breathing harder than at any time he could remember in his life. Apparently he wasn't in as good a shape as he had thought he was. Obviously he'd spent too many hours in the lab and not enough outside over the last few years.

They crested a slight rise and entered a stand of pine trees. Ahead he could see the lights of the security perimeter of the base. There seemed to be a lot more activity going on than when they'd left, which made sense after that wild lightning storm.

Stephanie glanced at her watch. "We're at five minutes."

She knew the tsunami possibilities, probably remembered the training and time line as well as he did. He wanted to stop and listen for the waves, but he was panting too hard to be able to hear the pounding surf behind them. He just hoped the surf was still pounding. As long as it was, they were fine. It was when the surf went silent that they were in trouble, especially since at the moment they weren't much more than twenty meters above ocean level. Not nearly enough to be safe from even a moderate-sized wave.

Forty-five more seconds of running and they were

at the guardhouse and the gate in the high chain-link fence surrounding the facility.

"Wild storm, huh, Dr. Downer!" Craig shouted as they ran up. "You two caught in it?"

Sergeant Craig was one of the two guards stationed on the gate. He wore the standard Union uniform and carried only a pistol sidearm. But Hank knew that two steps away inside the guardhouse was a Bulldog support rifle and other weapons. These days, Union soldiers didn't go far without full equipment, even on a duty like this one, guarding a bunch of civilian scientists.

Hank liked Craig. Actually, Hank liked most of the Union troops stationed at the facility. A good bunch of soldiers. He was glad they were there, even though some of the scientists complained of too much military.

"We were," Stephanie said.

"Better button everything up," Hank said after taking a deep breath. "We thought we saw something damn big come down out there in the ocean."

"Shit," Craig said. "You think it might be a tidal wave?"

"Good chance of one," Stephanie said. "And it might be a big one. On the other hand, might not be one at all."

"Let's not take any chances," Craig said. He took two steps and punched an alarm button on the side of the guardhouse, then quickly keyed in a code on an electronics pad.

Instantly the lights around the forest came up to bright intensity, pushing away any shadows. A siren sounded, three quick blasts, then pause, then three quick blasts, then pause again.

Then it repeated.

That pattern meant take cover inside, the facility was about to lock down.

"Get inside, you two!" Craig shouted.

Stephanie took Hank's hand and pulled him at a run down the paved driveway toward the underground complex. The complex only had one entrance, a ten-meter-high set of heavy, vaultlike blast doors on the outside and another set thirty paces inside the tunnel. Hank knew that the thirty paces between the doors were solid concrete all the way over and under the entire facility. It had been made earthquake- and nuclear-attack proof, and if a big wave was coming they were going to need the protection.

Hank actually managed to keep up with Stephanie, sand in his boots and all. A few dozen other civilians and guards were running ahead of them.

It wasn't until they were almost to the first door that Hank heard what he'd been very afraid they were going to hear.

Silence.

The continuous background sound of the ocean surf had completely stopped, and a calm had settled over everything. Even the wind seemed to have suddenly stopped.

He looked back as he ran. Craig and the other guard were a good fifty steps behind them, coming at full speed toward them, their rifles held in front of them.

"It's coming!" Stephanie shouted.

Ahead of them the rest cleared the first door and kept running.

Hank and Stephanie ran side by side, making the last twenty steps through the huge metal outer doors of the facility in what seemed like a silent eternity. Then they went past the equally heavy inner doors.

In front of them stretched another twenty meters of well-lit tunnel, with a guard post on the wall to the right. Beyond all that was the facility, one of the Union's top research and development establishments, staffed with hundreds of its elite.

"Get those doors shut!" Craig yelled behind them as he reached the outer door.

"Be ready to close the inner ones, too!" Hank called to the guards inside the tunnel guardhouse.

Slowly, almost too slowly it seemed, the giant doors swung shut as outside a roar seemed to be building. It sounded as if the entire planet was suddenly becoming angry and growling low and deep at them.

Craig and the other guard cleared the inner door just as it started to close.

But Hank wasn't so sure the outer door was going to get shut in time.

Seconds seemed to tick by as the rumbling, crashing sounds outside grew and grew.

The huge doors got closer and closer together, then finally banged shut with a finality that sent a wave of relief through him.

A few seconds later the inner doors did the same.

Safe.

He let out a deep breath. Stephanie did the same.

"Lock everything down!" Sergeant Craig ordered.

"Locked down," the guard replied.

Then, as if the ground around them was angry that the first huge wave couldn't get to them, everything shook.

Dust sifted down from the concrete above them, but nothing broke loose.

Not even the lights flickered.

And then it was over. The first wave had passed. If it were a normal tsunami, there would be others following.

Silence filled the tunnel.

Finally Craig turned to Hank and Stephanie. "Thanks, Docs," he said. "I have a sneaking hunch you two just saved our lives. I owe you."

Hank looked at the huge metal doors filling the end of the tunnel and laughed. "Don't mention it," he said. "I'm just glad to be standing right here."

Again the ground shook around them as a second wave pounded the outside entrance.

2

D r. Stephanie Peters stood next to Sergeant
Phoebe Malone in the entrance tunnel as the
huge metal inner doors to the facility opened.
The antiseptic smell of the facility's processed air was
instantly overwhelmed by the smell of the sea—a
thick, almost fishlike odor that Stephanie knew at once
didn't bode well for what they would find outside.
Normally, the grounds smelled of pine trees from the
surrounding forests. The ocean scent was usually a
distant, background presence, never this strong.

The exterior security cameras had all been
knocked out last night, so they had no idea what they

would find this morning. Considering the strength and number of those monitors, it wasn't likely to be anything good.

The wide concrete tunnel between the two blast doors looked undamaged, so Malone turned and shouted, "Open up the outer doors!" Her deep voice echoed down the tunnel.

Phoebe Malone was one of the toughest, take-no-prisoners members of the Union guard assigned to the facility. She had chopped short black hair, a strong body, and a deep voice that Stephanie thought could take down an enemy soldier with a yell. She had never seen Malone in anything but standard-issue uniform and doubted the woman even had any other clothing.

Stephanie had at times made various friendly overtures toward Malone, but the sergeant had always brushed her off. Phoebe Malone was obviously one of those military types who would rather be on the front lines than wet-nursing a group of scientists.

Slowly the outer doors opened with a grinding and snapping sound. Usually they made no sound at all. Another bad sign.

Finally, the two massive steel doors were open wide enough to let one of the facility's vehicles out, and Sergeant Malone shouted for them to halt.

The early-morning light wasn't bright coming through the opening, but then the Maw's light never seemed that bright to Stephanie, even on a clear day when it was directly overhead. This early in the morning it seemed more like an artificial light was on outside the doors.

"Stopped, Sergeant," the guard in charge of the door shouted.

"Stand by," Malone shouted back, then she and two of her men went through, picking their way in the debris that the doors had pushed back.

Stephanie and three of the other medical staff followed, growing more and more shocked at what she saw. Logs, sand, and brush filled the area in front of the big doors. And everything was wet.

The ocean was now visible down the hill, the gentle blue waves gleaming in the morning light. But before last night it had been impossible to see the water from the door of the facility. There had been a thick forest and a fairly large sand dune between the entrance and the beach. The same dune she and Hank had scrambled up in trying to outrun the storm.

Now every tree lay smashed flat, many sticking half out of the sand. And the top of the dune was completely gone, with massive streambeds cut through it from the water flowing back to the sea.

The path she and Hank had used to run back from the beach was nowhere to be seen. In fact, the more she looked, the more she realized that the beach was farther inland by a good distance.

The five-meter-tall chain-link fence that had surrounded the facility entrance was smashed in places and gone in others. The main guardhouse where Craig and the other guard had been stationed last night was nowhere in sight. She and Hank really had saved their lives.

Sergeant Malone stood to one side and surveyed

the damage as a couple of the other scientists joined Stephanie and the medical crew.

She turned, hoping to see Hank among them. He wasn't. It was Dr. Richard Lee, the Union's foremost expert in computers, and Dr. Jeff "Chop" Edaro, its top man in molecular physics. If there was a rule of the universe he didn't understand, no one did.

Lee was tall, with gentle hands and a smile that could disarm a Neo-Soviet soldier. Edaro, "Chop" to his friends, was intense and just as friendly. And not once would he tell them how he got the nickname. Lee claimed he knew, but would only say it came from a trip the two had taken with a couple of other friends to Phoenix. And that the name had cost him money.

Though the facility's researchers were often working on diverse projects, they'd already spent months on end together underground, and so tended to get to know one another. Sort of like a high school where no one ever went home.

Stephanie looked back down the tunnel and sighed. She and Hank had passed the last few hours since barely escaping the wave curled up together in his bed. When she tried to rouse him this morning, he said he wanted to get another hour of sleep before going to the lab. She'd thought he was kidding. Sometimes she didn't know how he could stay so focused on his work.

"Holy smokes," Edaro said as he came up to her. "You two were really lucky. How many waves hit here?" Even then he was playing with a little golf ball he always seemed to be carrying, flipping it softly into the air. Stephanie had found it a little disconcerting at first, but now she was used to it.

"Seismographs say four big ones and two smaller," Lee said, coming out the door to look up the hill above them. "Looks like it went another twenty meters above the entrance. Too bad the exterior monitors didn't survive. It would have been interesting to watch."

Stephanie turned to stare at him. "Interesting for who?"

"What about Dustin Cove?" Edaro asked, still tossing that little ball into the air and catching it.

"There's been no word from the troops and civilians in Dustin Cove," Sergeant Malone said coldly. "Rescue crews are coming over from Portland to help out there. Our orders are to stay put and get back to normal as quickly as possible."

Stephanie felt as if someone had kicked her in the stomach. She and Hank had worried last night about their friends living in Dustin Cove, knowing all they could do was hope the tsunami wouldn't be as bad as they feared. Now, of course, Stephanie could see the truth. The waves had been worse.

Bigger than she had feared they would be.

She glanced in the direction of Dustin Cove, but couldn't see it. From the looks of the destruction around them, they'd be lucky to find any of its inhabitants alive.

"Any idea how far up and down the coast the destruction was?" Dr. Lee asked.

"None," Malone said.

A private named Skinner came out the door,

glanced around, then approached Malone. "Sergeant, we have satellite confirmation on the island."

"Island?" Stephanie asked.

Malone nodded. "Seems an island about two kilometers wide and six kilometers long appeared out there about fifteen kilometers offshore. That's what caused all this."

"You're kidding, right?" Lee asked. "Volcanic?"

"I have no idea," Malone said. "That's up to you scientists to decide."

Stephanie stood there speechless as Malone set up guard stations. Then she picked her way through some fallen logs and headed back inside, Skinner close behind her.

Stephanie glanced up at the angry blue-white eye of the Maw that lit the sky, then turned and stared out over the sea, trying to catch a glimpse of what she and Hank had seen come out of the sky last night. No rock that big could have fallen from orbit and not been a terminal event, a cataclysmic happening so large as to destroy all living things on Earth.

Yet now they were saying something had come down out there, something big. And she and Hank had glimpsed it in the midst of the horrific lightning storm. Something big and dark, slowly lowering itself into the ocean.

"I have a sneaking hunch," Lee said, "that pretty soon some of us are going to find out exactly what landed out there."

"You mean an expedition?" Stephanie asked.

"Count me in," Edaro said, pocketing his golf ball

and smiling. "Anything to get me out of that lab and into the air and light."

"Such as it is," Lee said, motioning up at the Maw, which had replaced the vanished sun.

"Yeah, still better than fluorescent anytime," Edaro said.

Stephanie wasn't so sure, but the idea of seeing what was out there intrigued her, too. What could possibly have landed softly in the ocean, yet was six kilometers long and two kilometers wide? Malone had said she didn't know, but Stephanie didn't think anyone could answer that question right now because none of this made sense.

She looked back at the ocean, which looked slate blue and unusually calm, almost as if trying to make up for smashing everything last night.

Almost.

Time: 9:15 A.M. Pacific Time
7 hours, 44 minutes after Arrival

Major Frank Lancaster sat behind his desk, waiting, his hands clasped calmly on top of the appointment book in front of him as the six staff scientists filed into his office and took seats. He'd slept very little last night, having been awakened at one-thirty by the shutdown sirens. He could feel that his eyes needed drops, as if there were tiny grains of sand stuck under the lids. His hands were shaking slightly from too much caffeine, but he knew that soon he would need even more coffee. This was going to be a long day.

It was lucky Downer and Peters had seen what

they'd seen last night from the beach and that Sergeant Craig had acted quickly in buttoning up the facility. Otherwise, they'd be facing a lot more damage today than a lost guardhouse and some missing fence.

On the corner of his desk were the first reports coming in from Dustin Cove. It looked like they were going to lose a lot of good people there. But at the moment he couldn't let himself think too much about that. He had other business that was even more pressing.

He nodded to the six scientists, then as Sergeant Malone came in, he said, "Close the door please."

She did as he asked, then took up a position standing against the wall, hands behind her back.

The six people sitting in front of him were all experts in one field or another, some of the best the Union had. He didn't especially like the idea of sending them away from the safety of this hill. But he had his orders. He looked at each one, nodding. He had handpicked them for their specialties, knowing they would make the best team available on such short notice.

Stephanie Peters was the youngest at thirty, an M.D. stationed at the facility to do research into the effects of radiation on the human body. After the Russians nuked Seattle and Detroit, the Union had stepped up its work to fight the effects of radiation. Since Seattle was so close, she was here.

Hank Downer, thirty-eight, was a physicist and also the world's leading expert in laser technology and related areas.

Richard Lee, thirty-nine, was a master of computer languages and programming. There didn't seem to be any computer he didn't like. Actually, there didn't seem to be much in life he didn't like, since he always seemed to be smiling.

Kelly Bogle, thirty-nine, was a Ph.D. in angular physics, among other degrees. His work here was cutting-edge stuff concerning orbital weapons.

Jeff Edaro, forty, was a molecular physicist, currently researching nanotechnology. He was a golf nut beyond all reason. He and Dr. Bogle would head inland to a local course at every possible opportunity. Even now Edaro had that golf ball in one hand, nervously rolling it from finger to finger.

Bradly Stanton, thirty-seven, was also an expert in computers and had a dozen degrees. He had been part of the initial secret investigation of the Cache and had played a big role in successfully decoding the information they had found there.

The Cache was a planetoid that a Union explorer ship had discovered in the course of countless missions sent out to learn more about the Maelstrom. Alien ruins as well as the remains of high technology were scattered over the planetoid, including tapes showing how a race the Union had named the Pharons had destroyed the inhabitants of the planetoid. Lancaster had seen some of those decoded images, and it wasn't a pretty sight. Here at the facility, Stanton was still studying and testing some of the alien technology they'd discovered in the Cache.

"Good morning, ladies and gentlemen," Lancaster said, starting the meeting. "By now you've heard about Dustin Cove, and I'm sorry to report that I still have nothing more on the situation there."

All of them nodded, but no one spoke.

There was nothing they could say. Over the last six years, since the war had started and the planet had been sucked into this strange space, death had become a way of life for almost everyone. Many had lost friends and family when Detroit and Seattle were annihilated. Others in the mass destruction caused by the Earth being sucked into the Maelstrom. By and large, the scientists in the facility were a tough lot. They would keep working no matter what happened.

He quickly moved on. "But that is not the only reason I called you here this morning. I've been asked to assign you to a new, temporary duty. I've talked to a couple of you over the last hour, so some of you already know something about what's involved."

"We're going exploring, huh?" Lee asked, smiling.

Lancaster nodded. "Exactly. You six, along with a squad of Union troops led by Sergeant Malone, will be heading out for the island that's suddenly appeared out there in exactly fifteen minutes."

"What's the rush?" Bogle asked, his dark eyes focused and penetrating.

Lancaster smiled thinly. "Trust me, Dr. Bogle, we want to get there ahead of the Neo-Soviets. And since it's only fifteen kilometers off our coast, I think that's possible. We need to secure this island and determine whether or not it's a threat to our security."

"Okay," Downer said, "you want to claim a new island for the Union. What do you need us for?"

"I've got orbital photos coming in," Lancaster said, "good close-up stuff you can use. From what the brass in Colorado are telling me, we've got an alien city on that thing. An abandoned one."

All six of them started talking at once, and he let them go on for a moment. Stanton looked as if he'd seen a ghost. Since he'd been on the team studying the Cache planetoid, Lancaster didn't blame him.

Lancaster held up his hands for silence, and they all quieted down quickly, except Bogle.

"How can you know what's on that island is alien?" Bogle demanded.

Lancaster shrugged. "I'm only repeating to you what I got from up the ranks. The experts, from orbital pictures, seem to think there's something alien on that island."

"An entire city?" Lee asked.

"Covering the whole thing," Lancaster said. "At least that's what they told me."

Silence settled over the room as the words sank in. Clearly they were stunned, as he had been. After giving them a few more long seconds, he went on. "At this point, you now know as much as I do about that hunk of land out there. Except for how it got there."

"It came down from somewhere," Downer said. "Stephanie and I saw it."

"Well, you're half-right," Lancaster said. "They tell me it actually did settle a half kilometer or so into the ocean. But it didn't come from space. None of the orbital battle stations picked it up. And none of our early-warning systems caught it."

"How can that be?" Stanton asked. "I thought nothing could get through our air-defense system these days."

"I don't think anything this size could," Lancaster said, looking at Stanton. "And this island didn't come through our defense system. It simply materialized out there."

"Major, you lost me about two sentences back," Bogle said, shaking his head. "An alien city, covering an island, materializes out of thin air. Right? I think the boys in Colorado have been watching too much late-night television."

"I agree," Edaro said, pocketing his golf ball for a change. "We've all seen and heard of some pretty strange things over the last few years, but this tops them all."

Lancaster nodded. It was hard to believe, but he'd seen the tape decoded from the Cache. Alien life existed out there, and not so far from Earth. But except for Stanton, the rest of them had no reason to be privy to that information.

Lee looked directly at Lancaster, his eyes intense, his smile gone. "So where did this island come from? Anyone have any idea?"

"Your guess is as good as mine," Lancaster said.

"Let's assume what the major is telling us is true," Edaro said. "And an island just appeared out there. By my way of thinking, what can appear can disappear."

"Damn," Hank Downer said. "That means we haven't the slightest idea how long it's going to remain out there. It might be a permanent new island, or it may be gone in six hours. No way of knowing. Right?"

Lancaster could only shrug. He had no idea.

Suddenly some of the excitement he'd seen on

their faces a moment before had turned to concern. And a touch of fear. He didn't blame them in the slightest. He didn't much like the idea of sending them somewhere that might vanish at any minute, killing them or carrying them off to who knew where. But he had no choice. The Union needed to secure this alien thing, and it was crucial to find out what those ruins might reveal. It was a matter of national security.

"So the mission is simple," he said, again breaking the silence that filled the room like a heavy weight. "You go in, you find out what you can, you collect any tech or specimens for further study, and then get the hell out."

"Just like we did on the Cache," Stanton said softly.

Bogle looked at him and frowned, but didn't ask. For the moment, Lancaster was glad for that.

"Which is why you want us along for the ride instead of just sending in troops," Edaro said. "You need people who can identify whether alien technology is important or not on sight. Am I right?"

Stanton snorted. "If there is truly alien stuff out there, we'll be lucky even to understand what it was, let alone what we could use it for."

"Do what you can," Lancaster said. "I'll have a second squad standing by here, in case you need backup for anything. We'll also be monitoring your every move, ready to pull you out quickly."

Again that didn't make any of their faces any lighter.

"Okay," he said, taking a deep breath. This next part was the most important, since he couldn't, technically, give these people orders. "None of us knows very much about this island, but in a few minutes I'll show you the satellite photos and then you'll be as up to speed as anyone can be on this thing right now. You're scientists and not soldiers, but this is your chance to do something that will directly safeguard the Union as well as benefit in other ways we can't imagine right now every single person living in it. Not many of us get the chance to do something like that in our lifetimes."

Dr. Peters glanced at Dr. Downer, then turned back to Lancaster. "I assume you've included me as the team's medic as well as for my work on radiation. I'll go."

Lancaster nodded. "Thank you."

"Count me in," Edaro said, the golf ball back in his hand.

"Me too," Lee said, smiling again. "I've done stupider things, I suppose."

"I'm in, too, Major," Bogle said. "Like you said, it's the chance of a lifetime."

"Crazy is what they call this," Stanton said. "I know. I've done it before."

"Does that mean you'll go, too, Dr. Stanton?" Lancaster asked.

"I can't pass up the opportunity to learn more about the dangers that might await us in the Maelstrom," Stanton said, "but I can't say I like the idea."

Lancaster nodded. "Thank you all. We'll give you full support. Sergeant Malone is in charge of the overall military mission, but I have ordered her to follow your lead while on the island as much as possible. If push comes to shove, however, she is in command. Period. No questions asked. Everyone understand that?"

"We do," Stanton said. The rest nodded.

"Dr. Downer, I'd like to ask you to head up the civilian team—unless anyone has an objection."

Again silence. From Lancaster's experience with any group of scientists, this was the quietest he'd ever heard.

He glanced at his watch. "Okay, people, food, weapons, and equipment are already being loaded on the transport. Dress warm and be outside in ten minutes."

"How long will we be out there?" Downer asked as they stood.

"Plan on twenty-four hours. It might go forty-eight. But if it does, we'll pull you back here at that time for a debriefing."

"If the place stays around that long," Bogle said.

"How about we don't dwell on that point too long?" Stanton asked.

Bogle only shrugged.

"Good luck, people," Lancaster said, then watched them file out. After they were gone he glanced over at the waiting Sergeant Malone, who was watching him with a serious expression.

"Keep them alive and bring them back, Malone. That's an order."

She snapped to a salute. "Yes, sir. I will, sir."

He returned her salute, then said, "I have to show you something before you go." He clicked a button on the face of his desk, and a screen lowered into place out of the ceiling. "It won't take very long."

Malone stepped in closer at a gesture from him.

"There are aliens in the Maelstrom, Sergeant. We know that for a fact. You need to understand that fully to do your job out there."

"I'll do my job, sir," Malone said.

He nodded. "I know you will. The island looks abandoned, but you've got to be prepared for the worst. It's still possible that you might encounter alien life-forms out there. You need to be prepared for that mentally and psychologically."

The sergeant only nodded slowly.

"This is classified information," he said. "But since you're headed for what looks to be an alien city, you've got a need to know."

She nodded and faced the screen, still not speaking.

Lancaster tapped a button to start the five-minute playback of the decoded images brought back from the Cache, then watched it again with the sergeant.

And again, the images of what those aliens were and what they could do to another race chilled him to his very core.

4

C old last night. Hot this morning. Hank Downer was starting to understand that the weather along the Pacific Coast was never something to take for granted. He wondered if it had been this unpredictable before the Maelstrom had taken Earth. He was from northern New York State, where the weather consisted of four distinct seasons. You could predict it, almost set your watch by it. He hadn't been this far west in years, let alone lived there before his assignment to the research facility. He figured it would take him a few more years to get used to the weather, no doubt about it.

Hank stood next to Stephanie on the sand-covered landing pad to the right of the tunnel entrance and watched the final loading of equipment and supplies into the Hydra transport. Before last night's tsunami, this pad had been well-swept concrete, surrounded by tall pine trees on three sides. This morning they'd found it covered with fallen trees and big piles of sand, both of which had been mostly removed. There hadn't been enough time to sweep all the sand clear yet.

He wore a thick ski parka and carried a raincoat over his arm. He had a stocking cap stuffed in one pocket of the coat and gloves in the other. There was no chance he was going to get as cold out there on that island as he had gotten last night. But at the moment, in the morning light from the Maw coming up over the coast range, the parka was too much.

Way too much.

He could feel a small trickle of sweat start on his neck. And the parka made the morning air feel twice as warm, since it blocked the onshore breeze. He unzipped the coat and held it open, letting the ocean air cool him a little.

In a small pack on his back he had a change of pants, underwear, two shirts, and three pair of socks. With luck that would be enough to keep him dry and warm until they were pulled out. He'd brought no food, since that was supposed to be supplied. He just hoped they remembered it.

Stephanie squeezed his hand, then let go to check something in her pack. She had also dressed warmly, with a thick coat, a raincoat, gloves, hat, and enough extra clothing to make it an extra day. She carried a small medical bag, since she would be the sole medical doctor on this crazy expedition.

And crazy described it completely. He didn't know what he thought of Major Lancaster's announcement that there was an alien city on an island off the coast of Oregon. It had been so matter-of-fact. And when a man like Lancaster could accept an alien city on an island as a hard fact, it was difficult to doubt it was true. Over the last few years Hank had speculated, as many people did, that other races might exist in the Maelstrom, but now that he was faced with it, the idea was disturbing.

He remembered back to the days when the Earth was suddenly seized and sucked into the Maelstrom. Then the outbreak of world war between the Union and the Neo-Soviets that had occurred almost simultaneously. It had seemed to him and millions of others that the world was coming to an end. They'd been torn from the solar system and hurled into a space that was truly a Maelstrom. At its center was the swirling Maw, which had replaced the sun in the sky.

No one seemed to have the slightest idea what had happened. He'd been teaching back at Georgetown University when everything went to hell. He spent two wild days trying to get away from the entire East Coast, along with hundreds of thousands of others, as panic overtook them. He had been convinced that Washington, D.C., was going to be bombed. Instead it was Seattle and Detroit that were hit.

After a few short weeks, he simply did what most everyone else did and went back to work. Over the next half year the Union tried to convince him that his mind would help the cause. They offered him a research position, and eventually he'd accepted. Now, almost six years later, most of the country had settled down, and life went on without a sun in the sky and stars at night. It amazed him how adaptive the human animal truly was.

Now he stood there, on the Oregon coast, about to board a transport to an unknown island that he'd actually seen appear in a freak lightning storm. How had he gotten so brave in just six years?

Behind him the voice of Dr. Bogle told him that the other four scientists were coming out of the tunnel to join them. Each wore coats like he and Stephanie, and each had a small pack over one shoulder. Nothing else. Everything else would be supplied.

Hank glanced back at the transport just as Sergeant Malone motioned with an abrupt wave for them to come aboard. The woman certainly didn't waste words. She was dressed in full Kevlon armor, right down to the helmet. In one hand she held a Bulldog support rifle as if it belonged there. Hank didn't think she was someone he would ever want to cross.

"That has to be a hot suit," Hank said, letting his parka flap open as they all climbed past her and inside the transport.

Malone said nothing, seemed not even to hear his comment. It wasn't any warmer inside, but it wasn't cooler either. And inside there was no ocean breeze.

"These transports are bigger than they look," Stanton said, glancing around the interior.

"Antigravity," Lee said. "Carries more."

Hank knew the Hydra transport used the new antigravity hover technology, but it was the first time he'd ever been in one. He'd done some work on a project that was trying to combine antitgravity with laser-power drives, but he'd been reassigned to another project early on and lost track of how the first one had turned out. Despite that, he had no doubt that laser-powered antigravity ships would soon be a reality.

The section of the Hydra they were in was obviously used for troop transport and nothing else, and there was no way to see the pilots from there. The hold was filled with rows of benchlike seats and smelled of grease and sweat. But Stanton was right—it did seem bigger inside than it looked outside. And there was no sign at all of the equipment and supplies that had been loaded just a few minutes earlier.

Eight Union troopers were already buckled in along one side, all of them in full gear, all with rifles standing beside them. Not one of the soldiers looked up as the scientists boarded. Hank thought maybe they were angry at being assigned to baby-sit a bunch of civilian scientists.

Sergeant Malone motioned to the row of seats opposite the combat squad. "Packs under the seats," she ordered.

Hank did as he was told, going to the end of the row and stowing his pack in the space under his seat. Then he buckled himself in, using both shoulder straps and lap belt. Sitting next to him, Stephanie did the same. A moment later Malone came over and checked them all, then went to take a seat at the head of her men.

"Ready," she said into her mouthpiece as she buckled her belts.

The door instantly slid shut with a sharp bang that startled Hank. The sound had a finality to it that he didn't like.

"You know," Stanton said, "I went to college purely to escape these kinds of things."

"Little did you know," Lee said, laughing.

Hank couldn't have agreed with Stanton more. He'd always been content to serve the Union cause with his mind, his inventions, his research, but this field duty seemed just a little out of line. He could have refused to accept the mission, but that just didn't seem like an option the way Major Lancaster had laid out the situation. It still seemed odd to be sending civilian scientists into the place first. Maybe Lancaster hadn't told them everything he knew because of security considerations.

The entire transport rumbled slightly and lifted off. Hank could feel the motion but no sense of acceleration. It was more a heaviness that told him they were moving even without the roaring sounds of engines. He thought to himself that this was a perfect transport vehicle for getting troops in and out of an area fairly silently.

Stephanie reached over and rested her hand on his, not so much out of fear, but for comfort. He had to admit he felt better with her close. And worried about her at the same time. There was just no telling what they were getting themselves into, and he wouldn't want anything ever to happen to her.

"It would be nice to see where we're going," Stanton said.

Bogle laughed. "Yeah, let's see this alien city."

Sergeant Malone reached up to a panel beside her seat and keyed in a few numbers. In response, a large, flat screen unrolled from the center of the roof. A moment later they could see the rolling waves below and ahead. It looked as if they were skimming just forty meters off the water, and moving fast.

"Thank you," Stanton said, but Malone gave no sign of acknowledgment.

Hank could tell that her squad was all business, too. He hoped it was only because they were taking every precaution, and not because they were expecting a fight.

For the next three minutes the waves below flashed past on the viewer. The color of the ocean seemed to have a richness about it today in the Maw's light. The blue seemed even deeper and the swells larger than normal.

Then Hank could see the large shape coming up on the horizon.

An island where there had been no island before this morning.

An impossible place jutting out of the ocean.

They all watched in silence, the transport vibrating only slightly around them as the island grew on the screen until they could make out the details.

Buildings.

Thousands of buildings,

And all very strange-shaped. There really was an alien city coming up in front of them.

"I'll be go to hell," Bogle said, softly.

"Trust me, folks," Stanton said. "That's an alien city out there."

"How would you know that?" Hank asked, not lifting his gaze from the viewscreen as the island and buildings grew in front of them. He was stunned at how beautiful the place was. From a distance the city seemed to be reaching for the sky.

Massive structures, higher than any tall building in New York, jutted into the air, many ending in seemingly sharp points. They seemed to be linked together at different levels by hundreds of sky bridges. It looked as if they were built entirely of something white and polished. Some of the taller buildings reflected the light from the Maw.

"Amazing," Stephanie said, her voice as awed as Hank was feeling. "Who built this?"

"That's what we're here to find out," Stanton said, sounding grim.

"Let's just hope they're not home," Edaro said.

That thought made Hank shudder.

The buildings looked shorter, more compact, closer to the water's edge. But Hank still couldn't tell exactly what scale he was looking at. What he thought was a low building still might be twenty stories tall when compared to the monster structures that reached up from the center of the island.

"Can we make a pass around it?" Hank asked.

"Please," Stanton said, agreeing.

Malone spoke into her mouthpiece softly, then nodded to Hank. "One revolution. The onboard cameras will photograph the island's entire coastline as we pass, feed it into the data received from orbit, and produce some basic maps for us to use."

"We'll need them," Stanton said.

Hank kept his attention focused on the images on the screen. The closer they got to the shore, the clearer one fact became. At first his mind tried to deny it, tried to keep the first image of beautiful towers reaching for the stars. But reality took over.

This city was alien.

And it was a ruin.

Many of the center buildings had once been taller; their perfectly smooth surfaces were ripped and jagged at the top. Others had somehow retained their pointed peaks. From the looks of the gaping holes in the smooth sides of many taller structures, there once had been hundreds more sky bridges linking them.

Hank guessed the canyons between the structures would be a mess, probably so ruined and full of rubble as to be impassable.

Between the lower buildings were smooth surfaces resembling streets, at least near the water's edge. From the transport's low angle skimming around the city, Hank couldn't see anything between the massive structures that filled the center. But if he had to bet, he would wager there were many, many levels under ground level in there.

Around the edge of the island he saw that there were buildings underwater as waves broke against and over many of them. The current shoreline wasn't the normal edge of this city. At one time it had been much wider and longer. They were only seeing what was left above the surface of the ocean.

"Hard to imagine this just transported here," Stephanie said. "What power could do that?"

"What power could snap the Earth into this weird space they call the Maelstrom?" Bogle asked.

Stephanie only nodded. But Hank could see what she had been driving at. Something, with a vast amount of power, had caused this city to appear here. Was it just random Maelstrom craziness or something more? That was a very important question to which he doubted they were going to find an answer in a day.

He tried to study some of the shorter structures near the edge of the water as the transport banked and began its circle. But it was the tall buildings crowding the center of the island that kept drawing his attention back.

In its day, the city must have been a glorious place, its shining surfaces caught and reflected by the light from an alien sun. What would it have been like to walk along one of its sky bridges with the teeming city below? Or stand in the canyons between them and just look up?

What kind of creatures had built this place?

What was it like when they'd walked those streets and lived in those buildings?

And what had happened to the beings that had lived here?

Suddenly he flashed on the image of New York City looking abandoned and ruined at some distant time in the future. And the thought made him shudder. At one time this island city had been home to unknown beings. More than likely, millions of them, from the size of it.

"The place looks like it was bombed," Edaro said.

"Or saw some serious fighting," Lee said. "Look at all the holes in those big buildings."

Hank glanced at Stanton, but couldn't read the odd look in his eyes. Stephanie, meanwhile, also sat staring, almost hypnotized.

The beings that could build such a beautiful city were obviously far ahead of humans in technology. And yet they had lost this city to something vast and destructive.

And very unknown.

5

Time: 9:52 A.M. Pacific Time
8 hours, 21 minutes after Arrival

Everyone inside the Hydra remained silent as it finished circling the ruined island city. Stephanie doubted she could have uttered a word even if she'd wanted to. The images of those exquisite towering structures, combined with the signs of violent destruction all around them, left her stunned. Her heart pounded with a mix of fear and excitement.

Part of her couldn't wait for the Hydra to land so they could investigate, find out who had built such a phenomenally beautiful place, but another part of her wanted to head back to the mainland right away.

Finally, the Hydra finished its circuit of the island

and banked hard to the left, heading toward what looked like an open area near the shore, maybe the end of a street. Stephanie noticed that unlike old-style planes, the Hydra's banking and other maneuvers weren't so much felt by its passengers as seen on the screen.

From where they were now, the elongated shape of the island was clear. The tallest buildings were in the center, with the smaller ones tapering off in all four directions. They were heading toward a spot about halfway up its coastline, where the island looked widest. It was the closest landing site to the larger buildings.

The transport stopped and hovered just offshore in the direction of the mainland, only twenty or so meters above the rolling waves. Across from Stephanie, Malone's squad members did a quick weapons check, initiated by a hand signal from the sergeant. They looked ready for anything.

Malone unbuckled her belts and went over to a panel near where Kelly Bogle sat. She slid the panel aside and pulled out a Union military jacket and tossed it at Bogle. "Put this on and keep it on at all times."

"Why?" Bogle asked, holding up the jacket to look at it.

"Kevlon-lined," Malone said. "Might save your life."

Bogle stared at her for an instant, then nodded and unbuckled his belts as one by one Malone tossed each of them a jacket.

Stephanie held hers up. Surprisingly light for

something that could stop bullets, it was belted and had a dozen pockets. She quickly slipped out of her safety belts, pulled off her parka, and slipped into the Kevlon jacket, pulling it tight around her stomach with the belt.

Actually it was fairly comfortable. But she didn't like the idea that Malone thought they needed them. She sat back down, using her parka as padding.

Malone watched until they were all in their jackets, then asked, "Has anyone here ever fired a Pitbull assault rifle?" She pulled one out of a cabinet and held it up with one hand for all six civilians to see.

It was smaller than the rifles a couple of the Union troopers carried. Those were called Bulldogs, if Stephanie remembered right. The rest carried rifles just like the one the sergeant held up.

Stephanie kept her hand down since the only weapon she'd ever fired had been her uncle's deer rifle during her vacation in Colorado. But Hank, Bogle, Lee, and Edaro all raised their hands.

"Good," Malone said, handing Bogle the rifle she'd been holding. Then she handed one to each of the other men. "Ammunition is there." She pointed to another panel. "Take as much as you can comfortably carry with your other supplies. Keep the rifle and ammunition on you at all times."

The men nodded.

Stephanie watched as Hank expertly inspected his Pitbull. He had obviously handled one before, more than likely in his work developing laser rifles.

Then Sergeant Malone reached into the panel from which she'd gotten the rifles and pulled out two medium-sized pistols. "This is called a Pug," she said. She handed one to Stanton, then walked over and gave the other one to Stephanie.

Stephanie stared at the gun, feeling the metal ominously cold in her hand. The implications of their all needing guns and vests hadn't completely hit her until then. Even with the protection of a squad of soldiers, civilians had to be ready to defend themselves against something. She didn't want to ask what Sergeant Malone thought that "something" might be.

"You ever fired a pistol, Dr. Peters?"

Stephanie glanced up at her. "I have," she said, remembering back to a few hours on a target range. She might not be able to hit anything with it, but she could fire it.

"Good," Malone said. "Ammunition in here for those." She patted a different panel.

Then she looked at the men with the rifles. "Any of the rest of you want a pistol, too?"

Bogle and Lee held up their hands, and the sergeant gave them each one. Stephanie checked to make sure her Pug wasn't yet loaded, then put it in one of her jacket pockets. She could feel its weight resting against her leg like a bad omen.

The soldiers sat across from them, their faces neutral as they watched the view of the shoreline on the screen. They, too, were mute reminders of the real danger out there.

"Okay, Doctors," Malone said, stepping back to the head of the seating. "Here's what I want you to do. Get your packs on, tuck your extra coats into the belts on the Kevlon jackets, and get your ammunition. Do it now."

Stephanie stood, feeling unreal, as if this were a dream.

Beside her Hank worked quickly and efficiently, getting rifle ammunition.

She stuffed her rain slicker into her pack with some extra socks and tee shirts. The small pack went on over the Kevlon jacket, and she made sure it was pulled snug.

Then she used the sleeves of her ski parka to tie it around her waist, tucking the sleeves under the jacket belt to make sure it wouldn't come loose. She was just going to have to carry her medical bag. No other way.

Hank passed her some clips for the pistol. She took ten before the coat started to feel too heavy, then refused any more. If things got to the point where she had to fire that many rounds, Stephanie figured she'd be in such big trouble that having more ammunition would be a moot point.

When they were all ready, Malone said, "Buckle back up, we're going in. When we land my men will secure the area. After I give the all clear, and only then, you will exit the transport. Is that clear?"

"Very," Stanton said.

Stephanie nodded. She was glad Malone was being extra cautious, but this entire scenario was scaring her down to her toes. And she'd already been scared enough just at the idea of going into a ruined alien city.

"Good," the sergeant replied.

Stephanie sat down, the pack holding her away from the wall, the weight of the gun and ammunition heavy in her pockets. She managed to get the lap belt buckled as the transport dropped toward what looked to be an open street. The lower end of the street disappeared under the waves, and buildings lined the other two sides. It was the largest open area Stephanie had noticed in their short fly-around. This city was one compact place.

The screen rolled up and retracted back into the ceiling a moment before a bump indicated that the transport had landed on the island.

Stephanie's stomach clamped down tight, and she could feel her breathing getting quicker.

Instantly Malone's squad sprang into life.

Almost as one, they unbuckled their belts to a flurry of loud clicking sounds and headed for the hatch. Just as Malone reached it, it slid back and she went through, rifle ready.

Two by two, the other soldiers followed until they were all out, then the hatch slid shut with a clang that underscored the sudden silence.

"Not sure what scares me more," Stanton said finally, "the alien city or the troops guarding us."

Stephanie forced herself to take a deep breath. It didn't help much.

"I'm just glad they're along," Bogle said.

Stephanie was, too. Now that she was there she knew there was no way she'd set foot on this island without them.

Bogle undid his seat belt and stood.

She and Hank followed his example. Her hand

went to the pocket with the pistol, but she didn't pull it out to load it yet. It just felt good to know it was there.

"I'm damned glad they're along, too," Stanton said as he stood. "But they still scare me."

There was no sound coming from outside the Hydra.

Beside her Hank placed a clip into the rifle and snapped it home.

Stephanie looked at him. "You really think there might be trouble?"

He shrugged. "Anything could happen."

"Truer than you know," Stanton said grimly, loading his pistol.

Lee snapped a clip into his rifle, as did the others with rifles.

Hank glanced at her. "Better load a clip into the pistol just in case. You never know what we're going to find out there."

Stephanie nodded and pulled out the pistol. She slid a clip into place, made sure it was secure, then put it back into her pocket without pulling a shell into the chamber. Surprisingly, that simple action made her feel better. A little more confident, a little more relaxed.

Suddenly the door snapped open, startling them all. She felt as if her heart might explode right at that moment.

"Clear," came Sergeant Malone's deep voice from outside.

"Hey, Malone, you scared the hell out of us in here," Bogle grumbled.

Stephanie followed the rest to the hatch and then down the slight ramp to the ground, a concrete-looking surface that at one time might have been white, but was now stained with swirling black and brown. She wanted to bend down at once and inspect it, but she refrained and looked around instead.

The air smelled of sea, of dampness, and a faint rotting odor that Stephanie couldn't place.

The closest buildings must once have boasted smooth surfaces, but were now cracked and punched full of holes. She could see empty rooms inside, but nothing else. They were all about three stories tall, with flat roofs and no obvious doors or windows. The structures were linked together from the second story down, forming a continuous wall along both sides of the street. In fact, the street must have been a kind of a whitewashed gutter with three-story-high walls when it was new. Odd, Stephanie thought.

Very alien.

Then she noticed Hank looking up, and she followed his gaze, up into the unbelievably tall buildings at the center of the island. Even from this distance the buildings seemed to tower over them. Even the ones whose top portions had been destroyed were far taller than any Earth buildings. They gleamed in the white light of the Maw almost as if they'd been built to do so. In places they even seemed to sparkle.

Stephanie stood there in sheer awe, almost overwhelmed by both the beauty and the alienness of the place.

"I'd love to go inside one of those big towers," Hank said softly, so only she could hear.

"Maybe you will," she said. "Who knows?"

"This place is very strange," Lee said.

"Creepy," Edaro said, rolling the golf ball in the fingers of one hand.

She had to agree. Under her awe at seeing such an alien place was a sense of unease. And of danger.

She glanced around. Only three of the troopers were in sight, along with Malone. One stood guard while the other two unloaded the supplies and tested equipment. Stanton and Bogle had joined them to make sure the equipment didn't get damaged. Lee and Edaro were standing next to Hank, staring at the city spread out in front of them.

Suddenly two of the troopers emerged from a hole in a nearby building and gave Malone a signal.

The building they were in was the only one standing between two piles of rubble that had obviously once been similar structures. It faced out onto the wide street and looked solid. And defensible. At least to Stephanie's untrained eye.

"Okay, people," Malone said. "That building will be our headquarters. Let's get all this equipment in there and set up."

With one last glance at the tall, spirelike buildings at the center of the city, Stephanie turned and went to help.

The inside of the building was wide-open, a giant space bigger than any gymnasium she'd ever seen. It must have stretched clear across to the other street. A circular ramp led up to the next level, which was another seemingly large, open space with nothing in it. Both rooms smelled of dampness and decay. Both had debris piled in the corners and what looked like stains or watermarks on the walls.

Stephanie went over to one wall, pulled out a flashlight, and lightly ran her free hand over the marks. The wall's surface was smooth and cold to the touch. The mark was a stain that didn't rub off and ran all the way around the room. Everything above the mark was a slightly different color. If she had to guess, she would bet it was a watermark, and that this area had once been underwater.

Then she noticed another watermark just slightly higher on the wall. And then a third up near the high ceiling when she shined her light up that way. This city had apparently been in other oceans or lakes at one time or another. But which oceans? Had it been to Earth before? Or were these marks from an alien ocean?

Did the island move around a lot?

A lot of questions that needed answering.

She glanced around the big, empty space. Hank and Bogle were investigating the pile of junk in one corner. Lee and Chop Edaro had gone up the ramp to the second floor to explore.

She looked around her. What could this place possibly have been used for? Had some creature lived here?

What did that creature look like?

What caused the hole in the wall?

Too many questions and no answers.

"People?" Sergeant Malone shouted to them. "We need to get the supplies unloaded."

After thirty minutes and a half dozen trips to the

shuttle, they had all the equipment moved, and Stephanie was sweating intensely. She desperately wanted to take off the Kevlon jacket. But she didn't.

She didn't even open it up.

And she noticed that none of the others did either.

6

The white light of the Maw glinted in Sergeant Phoebe Malone's eyes as she watched the transport lift off, moving almost silently east, back out over the ocean toward the mainland. She was actually relieved it was gone. They would soon be on the move and having to guard the craft would be a pointless waste of her resources. She and her squad were now alone on the island with the team of scientists. Nine, counting her, to guard six. She'd have liked to have more men but would do the job with what she had. Besides, a second transport would be standing by just off the island as backup if they needed it. This site would become their main base if the scientists gave word the island was stable enough to support it. But she knew that would take a few hours to accomplish.

One thing was for certain—within the next hour there would be good air-cover support around the island. That would ease her mind some, when it happened.

She clicked her uplink to the base. "Transport away."

"Copy."

The voice from the base was clearly audible in her earpiece. They would be monitoring the situation via feeds from three different helmet-mounted cameras. One in hers, one in Jenkins's helmet, and one in Cort's. The mission was as secure as it could be, considering the quick-response time.

She glanced around at the empty alien buildings, the slick walls, the high rooftops. She'd posted two of her men in the rubble of those buildings at points where they could see her. She couldn't see them, as it should be.

She turned and stepped back over the fallen wall, moving into the alien building. As she entered Private Jenkins was spreading the photomaps out on the camp table he had set up. Jenkins was a big, red-haired man. All her men were good, but he was one of the best. She relied on him and the dark-haired, slow-talking Private Cort.

The maps had come from photographs supplied by orbiting battle platforms combined with images from the Hydra's low pass around the island on the way in. The maps looked almost three-dimensional, but she could tell that a lot of information was missing. Many of the streets between the tall buildings had photographed so dark that there was no telling what was down in there.

Or inside any of the thousands of buildings. Buildings seemingly without doors or windows. She knew the orifices had to be there, but so far she couldn't figure out where. A problem to be solved later.

The images of aliens cutting down whoever had been their foe flashed through her mind. She was glad the major had briefed her with that tape from the Cache. Those creatures had been damned ugly, grotesque but well armed. The mummylike look of their legs and arms had spooked her some, but she hadn't let on to Major Lancaster. Just seeing factual evidence of the existence of aliens in the universe had shocked her to her core. The fact that they wore golden armor and seemed to be walking dead was almost beyond her.

But at least she hadn't been sent in without a warning that they might run into just about anything. If it were up to her, though, that wouldn't happen. The last thing she wanted was to run into a living, breathing creature from God knew where. Especially one like those in the briefing tape.

The six civilians gathered around the map and began to discuss which direction they should head, what area would offer the best prospects for investigation.

She ignored them and turned to make sure the

rest of the equipment was getting set up. The first thing running when they got off the transport had been a proximity motion detector, and her people as well as the scientists had been given spotting devices to wear in their vests. That detector would tell them instantly if anything strange moved within a hundred meters of this location in all directions. It would also give her the exact location of any of her squad or the civilians, including what level of the buildings they were on.

The detector could sense heat and motion from bodies through walls and ground. So far they hadn't run into anything it couldn't penetrate. And luckily the alien walls weren't blocking it. Her men showed up on the screen as blue dots, the civilians as green dots. Anything else moving would show red. The device would be operational for as long as they were in the alien city.

Radiation detectors and the energy sensors were coming on-line as she moved over to the small stand holding the compact equipment.

"Report," she ordered.

Private Waters, the youngest member of her squad, didn't even look up. "Low radiation," he said. "But I have a faint energy source."

"What?" Dr. Stanton asked, turning from the map and crossing the twenty steps to where she stood.

"Pinpoint it, Private," Malone said, ignoring the rest of the civilians also gathering around.

"Hard to do exactly, Sergeant," Waters said, working the equipment. "The signal I'm getting from the energy source is weak. Shielded maybe. I'd say at least a kilometer directly inland from here. Underground maybe. Sorry, just can't tell."

"Human or alien?"

"Can't tell that either, Sergeant."

"What kind of energy?" Stanton asked.

Waters only shrugged.

"We need to get closer," Malone said. "Is that what you're telling me, Private?"

"Exactly," Waters said.

Dr. Downer glanced around at the others. "I say we check it out."

The rest just nodded.

Malone agreed with them, but for different reasons. Their job was to investigate scientifically any technology or phenomena that showed up here. Her job was multipronged: to protect the civilian team so they could do their work but also to scout the island to learn if it was a threat to the Union. Finally, she was to secure it for the Union to keep it away from the Neo-Soviets. If the energy source the private had pinpointed was the only one on the island, that's where she wanted to go, too.

"Pack it up," she ordered Waters and Jenkins. "We move out in five minutes."

She strode over to the maps, the scientists following like dogs at her heels. She stared at the images spread out on the table for a moment. She had studied maps for years, knew what she could get out of them and what information she couldn't trust from them.

She ran her finger along the wide street or boulevard that fronted this building. "We go inland along this route for as far as we can. When it's blocked, we move to the south and keep heading inland."

She turned to Downer. "That suit your purposes?"

"Perfectly," he said, nodding.

"Good. Make sure your weapons are loaded and be ready and outside in four minutes."

She turned and went back toward the hole in the wall leading to the street. On the way she keyed her commlink to the frequency only her men shared. "We move inland in four minutes. Gather up at the shuttle site. I'm going to want two-by-two cover movement on point down both sides of the street. Vasquez, Hawk, you two will keep a good distance behind the main group. Standard civilian-cover formation otherwise."

Then she climbed through the hole in the wall and out into the light. The civilians followed her. A moment later Private Jenkins came out carrying the equipment on his back. Waters followed him with the maps and the motion detector, which was still on and functioning. They'd all brought some extra equipment along for the mission. A little of everything, not really knowing what they might run into.

She nodded at Waters, then glanced around at the scientists. A couple of them held their rifles as if they knew how to use them. She made a mental note of whom, then looked inland along the street. The alienness of it all didn't much impress her. She had fought in her share of strange environments over the past years. This was just one more.

She watched as Privates Cort and Marva took up

positions along the right side of the street and Harden and Raynor along the left.

Hawk and Vasquez fanned out back toward the water.

Jenkins and Waters flanked the scientists.

"Everyone spread out at least four meters apart," she told the scientists. "Keep that distance as long as we're moving. One stops, we all stop. Understand?"

They all nodded.

"Good. Stay behind me. Jenkins, make damn sure that motion detector is working. I want to know if something's coming at us."

"Got it, Sarge," Jenkins said.

She turned and looked toward the tall, alien buildings filling the center of the island. "Move out," she said into her commlink. She knew that as soon as they cleared the area another shuttle would land and secure this position as a base.

Malone watched as her four lead men went ahead about a hundred paces, then she walked out into the middle of the alien street, hands poised on her Bulldog rifle. She began to move forward, pace slow and sure, timed to keep the same exact distance behind her point men.

She knew the camera in her helmet was sending back pictures that would be an important source of information later on. She was going to do this right and by the book. And they were all going to get out of here alive, if she had anything to say about it.

Behind her she could hear her charges fall into position. But her gaze didn't waver, scanning the openings of every building, the shadows cast by the ruins, the open spaces between the structures running to the left and right.

Like a well-oiled machine, the little brigade moved forward, deeper into the tall buildings of the ruined alien city.

7

Time: 10:42 A.M. Pacific Time
9 hours, 11 minutes after Arrival

 ilence.
       Deadness.
       Ruin.
Alienness.

All of it ate at Hank, dug at his nerves, twisted his stomach. But it was the silence that bothered him most. It pressed down on them from the tall white buildings like a heavy, invisible weight. They were far enough into the smooth canyons between the buildings that not even the wind seemed to blow there. And the light from the Maw only made it halfway down the featureless, ten-story walls, keeping them in shadow, increasing the feeling of pressure from the silence.

The smell of the ocean had faded, replaced now with a stale smell that at times Hank thought might almost choke him.

Moving ahead of the main group the four soldiers on point moved silently along the base of the buildings, one trooper on each side of the street giving cover while the other two ran forward, took up positions, and gave cover for the two following.

Hank watched as Malone walked slowly and carefully at the head of their group, her head moving back and forth, always vigilant, keeping perfect time and position with the point men.

He held the Pitbull assault rifle across his stomach, barrel facing down toward the smooth, stonelike surface of the street. It felt reassuring to hold it. And he was very glad he knew how to use it.

None of them had said a word for a long way as they moved up the street. They walked slowly, staring at everything they could see, not even the sound of their footsteps enough to break the quiet of the dead city.

Stephanie was to his right, four meters away, an intense look on her face as she gazed at the alien structures around them. She had one hand in her jacket pocket, probably gripping her pistol. Hank could tell that she, too, was unnerved by the silence and the dread of the unknown.

By the time Sergeant Malone called a halt, they had gone, by Hank's best guess, the equivalent of about six city blocks. But here in this city, there was

no such thing as a block. The street they were travel-
ing was cut at uneven intervals by other streets moving off at ninety-degree angles. And there seemed to
be no reason for the varying sizes, no patterns to them.
So far, in six human city blocks, they had gone
through eleven alien intersections.

Many of the buildings they passed had large holes
in their sides, as if the walls had simply collapsed.
Hank could see no sign that force had been used either
inward or outward on any of the damaged areas. The
walls looked like they had simply fallen, more than
likely from age. And falling down from age was far
better than seeing signs of battles. Evidence of fighting was the last thing he wanted to see.

He caught a glimpse through one of the holes as
they passed another large room, empty as the one they
had been in blocks back. Through another hole he
could see that the room's ceiling had collapsed, filling
the space with rubble. So far, besides the city itself,
there was no sign of alien technology or culture. Not a
stick of furniture or a piece of refuse on the street.
Nothing.

And no sign that anyone, or anything, had been
there since the city was abandoned.

Seemingly completely empty.

And silent.

"I can understand no windows," Stephanie said,
her voice firm. "But the fact that there are no doors is
starting to bother me."

"I've been thinking the same thing," Bogle said,
glancing back at the building they had just passed.

It bothered Hank, too. The fact of no obvious doors or windows had been eating at him since they first landed. Why build tall buildings if you couldn't see out of them? And they all had to have some way in and out.

"There has to be some hidden access in the walls somewhere," Stanton said.

"Sure," Stephanie said, "but how is it possible that not one of those access doors was left open when this place was abandoned? That seems to make no sense."

"And not one is visible from aging," Bogle added. "At least that I've seen."

Stephanie nodded.

"I grant you that it makes no sense to us humans," Stanton said. "But it's alien. It doesn't have to make sense to us. Just to the builders of this place."

Stanton was right, Hank thought. Nothing they'd seen so far made any sense. There was nothing here but silence and crumbling buildings. And that by itself felt alien.

Malone herded the group to the right side of the street and against one wall, then had Waters unpack the equipment.

Hank stood next to Stephanie, their backs against the wall as they watched. After about a minute the young soldier glanced up at Malone. "Energy source slightly stronger. Still a good three-quarters of a kilometer ahead, I would guess."

"Good, Private," she said. "Pack it back up and get ready to move again."

Hank glanced down the street that seemed to vanish into the dark shadows of the skyscrapers. The buildings near the water had been two and three stories. The ones around them were holding at around ten stories tall. But a short distance ahead the monster skyscrapers started. The energy source they had spotted was somewhere in there, among those monsters. But without doors in the buildings, how would they get to it?

Or escape if something was guarding it and came out after them?

He knew the door question had to be answered immediately, before they took one more step.

"Hang on, Sergeant," Hank said. "We need to get a couple questions answered before we get too deep into those larger buildings."

"Go ahead," she said.

He slipped his rifle over his shoulder and pointed at the rubble of a wall that had fallen, opening a path inside. "I think we should do a little digging through that rubble, see if we can find traces of how their doors worked. Or even if they had doors."

"Good idea," Stanton said, but he was moving his hands strangely almost as if he wasn't aware of it. The farther they had gotten into the city, the more nervous Stanton seemed to become.

Malone nodded, then spoke into her commlink. "Take up perimeter positions. Jenkins, Waters, make sure the insides of that building are clear."

She pointed at the building Hank had singled out, and the two soldiers nearest them headed toward it.

First they took up positions on either side of the hole, then went in quickly, low and guns ready.

Hank was almost holding his breath. He didn't know what he expected. He just hoped he didn't hear the sound of gunfire.

A long ten seconds later Malone nodded as she listened over her commlink, then said, "All clear."

That was the signal for the rest of them to pass through the opening. Inside, in the shadowed interior, was another large, high-ceilinged, empty room, with no indication at all of what it might have been used for.

Hank thought maybe it hadn't been used for anything. Maybe the aliens never came down to the street level. That was possible. But the only way to find out was to find doors.

Or ramps, or staircases, inside. There had been a ramp in the building near the ocean, but none that he could see in any of the bigger buildings they'd been passing in the last few blocks.

"Ask your men inside if they see any way to get up to the next floor."

Malone nodded and did as Hank had asked. In the meantime, Bogle, Edaro, Lee, and Stanton had all started to clear away some of the white wall material, kicking up a very light cloud of white dust as they worked.

"Nothing obvious," Malone said after a moment.

Stephanie shook her head. "There has to be a way up. In that building near the ocean the ramps were in the corner, moving up toward the center of the room. Right?"

"True," Hank said. "But I don't think that helps here."

Stephanie said nothing, clearly deep in thought.

"There's nothing in this stuff," Bogle said, standing back and slapping his hands together to remove the dust.

"Okay, so where are the access points?" Stephanie asked. "Whoever built these buildings must have had a way in and out."

"Don't assume anything," Stanton said, brushing off his hands on his pants as if trying to wipe off blood. "You're still laying human concepts on an alien culture. Trust me, I've run into this roadblock a hundred times over the last two years."

"Doing what?" Bogle asked.

"Later," Stanton said, waving off the question.

"So," Stephanie said, "you think they just built the buildings around themselves like tombs?"

"Possible," Stanton said.

"Wonderful," Bogle said. "We're exploring the equivalent of the Egyptian pyramids."

"It would be a true city of the dead," Lee said, smiling.

Stephanie smacked him on the arm. The rest only shook their heads at Lee's lame joke.

Hank, meanwhile, decided to go back out into the street. Malone stayed close to him, keeping her eye out around them.

He looked at the building. The idea that the ramps in the smaller structures were in the corners had him fascinated. Yet in the larger buildings like this one, there were no ramps. "Weird. Very weird."

Sergeant Malone stayed close, but said nothing.

"What are you looking for?" Stephanie asked, coming up beside him.

"Not sure," he said.

He forced himself to really look at what was different on the walls of the buildings. What appeared to be a clean, uniform surface along the side of the street actually had indicators where one building stopped and the next one started. Each roof was slightly higher or lower than the next. And many of the buildings filled a small alien city block from one side to the other. Hank thought it was possible that in some far distant past, these buildings had all been painted different colors. But now they were all worn down to their base material.

Yet every corner of every building was clearly marked. And just like humans putting arches over doors to mark where the entrances were, if the aliens put their entrances in the corners of the buildings, that would be the same idea.

Then he looked toward the center of the island and up at the edifices towering into the sky. Each sky bridge that connected two buildings always went from the corner of one structure to the corner of the other.

"Got it," he said.

"Got what?" Stephanie asked, following his gaze upward.

"The sky bridges," he said. "They're the answer."

Stephanie only looked more puzzled, staring up at the towers.

"Hey, everyone!" he called out to the others.

"Take a look at those sky bridges. Notice where they start on every building."

They all turned to look as he moved twenty steps back to the corner of the building, Malone and Stephanie following him. The sergeant also moved two of her men into a better position. She wasn't going to let any of her civilian chicks go unguarded for even a second. And that was fine by him.

Slowly, Hank ran his hand along what looked to be nothing more than a corner crack in the building's surface. But the crack didn't go any higher than three meters.

He tried to put his fingers into the crack to pry it open, but it was clear almost instantly that wasn't going to work.

"You think the doors might be on the corners?" Stephanie asked, as the other four scientists joined them.

In the background three of Malone's men moved into better cover positions while Malone herself stayed with the group of civilians.

"I'm convinced of it," Hank said. "How else would they know, on seemingly featureless buildings, where to go to enter or leave?"

"So how do we open one?" Lee asked.

"More than likely they were automatic," Bogle said. "And I doubt if there's energy running."

"Maybe," Stanton said. "Maybe not."

But Hank could tell Stanton didn't think much more of that idea than he did. Hank studied the wall up and down the crack, then down at the ground on the street. Of course a door wouldn't open out onto a street area. Bad design. So it had to open inward somehow.

"You know," Stephanie said, standing beside him, "we may be making this much harder than it really is. It's just a door after all."

"Be my guest," Hank said, smiling and stepping back out of her way.

Stephanie walked forward and simply pushed on the corner, directly inward.

Silently, both corner walls moved inward from three meters down, sliding back and open, exposing the empty room inside and one very startled soldier crouched against one wall, weapon ready.

Both doors seemed to be almost a meter thick, yet they moved as if they weighed nothing at all. Nor was there any sound of a motor or electrical device. More than likely they were just very well balanced doors.

Hank and the others applauded as Stephanie turned and smiled.

"Great thinking," Bogle said.

"Okay, one problem solved," Hank said. "How do you get out once you're inside?"

"Stay here," Stephanie said, then glanced at Sergeant Malone. "All right if I go in?"

Malone nodded. "Dr. Peters coming in," she said into her commlink. "Cover her."

Stephanie pulled a small flashlight out of her pocket and moved through the open doors.

As she cleared them, the two doors slowly closed behind her. Silently, seeming to glide on air, leaving only the faint crack in the edge of the building to show that the wall had ever moved.

"Amazing," Edaro said, staring at the door, holding his golf ball.

Hank instantly wanted to go after her, but then he remembered that two soldiers were in that large, empty room, and there was a hole in the wall about twenty paces away. But it still felt odd for her to disappear inside an alien door like that.

Odd and very dangerous.

"Stay with her, Jenkins," Malone said calmly into her commlink. Clearly she had felt the same thing.

A moment later the two doors swung inward, and Stephanie walked out, smiling at them, turning off the flashlight as she came.

"How the hell did you do that?" Bogle asked as the doors moved silently back into position as soon as Stephanie cleared them.

"Simple," she said. "There are indentations in the walls on either side with just enough of a lip to grab ahold of. Just pull slightly on one and they open right up. Come on, I'll show you."

Hank was impressed she had figured it out so fast.

"We're coming in, Jenkins," Malone said into her commlink, as Stephanie again turned and pushed on the door. "Everyone hold positions."

They all got out their flashlights and went through the door, which then closed behind them. Quickly Stephanie showed them where, on both sides of the corner, there were small indentations in the wall. Hank put his hand in one and could feel that what looked like an indentation was actually a lipped handle clearly meant to be grasped and pulled.

"Two problems down," Hank said. "One last one."

"Upstairs," Stanton said. "How do we get to the floors above this one?"

"Exactly," Hank said.

"Following the logic of the doors," Stanton said, "the stairs would also be in the corners in some fashion or another."

"Makes sense," Bogle said. "Same logic as putting the doors on the corners. Everyone always knows where they are."

"But why wouldn't ramps and stairs be built in permanently?" Stephanie asked, pointing her flashlight up at the top corners of the large space. "Doing anything else makes no sense."

"Thinking like a human instead of an alien again," Stanton said.

Stephanie shrugged.

Along with Lee and Edaro, Hank moved toward the center of the room, shining his light along the ceiling as he went, then ahead at the open space.

Sergeant Malone had four of her men scattered around the room, keeping a very close eye on them all.

As he walked he tried to imagine the room full of "stuff." Alien stuff, but the only image that kept coming to his mind was a department store. This space was big enough to house one, that was for sure.

"Hank, look at this," Lee said, pointing down with his beam of light.

Under the thin layer of dust that covered the floor were very faint lines, patterns, almost worn away by time. They were so faint, Hank was amazed that Lee had even noticed them in the dim light.

But now that he saw them, they were obvious. The line seemed to come from the corner with the door and go straight to the center of the room. Hank and Lee both followed the faint line, leaving footprints where no human had ever walked before across the vast area.

Just as they reached a certain point, the line split and turned and flowed into lines coming from both the left and the right. It seemed that the very center of the room had a circle on the floor.

Hank guessed that four lines on the floor led from the four corners to this circle.

Lee started to step over that line into the circle, but Hank instantly knew that he shouldn't do that. Not until they were ready.

"Wait," he said, holding Lee's arm.

The two of them took a few steps back and stopped.

Hank turned to where Sergeant Malone stood near the opening in the wall, watching them intently. The other four scientists were in one corner, talking about something on the wall and shining their lights at the ceiling.

"Sergeant, I think we might need a little extra cover here," Hank said.

Malone nodded. "Jenkins, Waters, join me. Cort, Marva, come inside and cover the opening in the wall. Everyone else hold positions."

All of them moved quickly to where Hank and Dr.

Lee stood. Stephanie and the others moved to join them also.

"What did you find?" Stanton asked.

Hank pointed at the very faint line in the dust on the floor with his light.

"What?" Stanton asked, looking down.

"I'll be go to hell," Bogle said. "Never would have seen it."

"What?" Stanton asked, still not seeing the line.

Bogle pointed along the line with his flashlight beam until Stanton suddenly grunted.

Hank looked at Malone. "Ready?"

"For what exactly?" the sergeant asked.

"More than likely nothing," Hank said.

"Or maybe a ramp," Lee said. Then, smiling at Hank, Lee stepped slowly over the center of the line into the faint circle and stopped.

Hank didn't know what he was expecting, but he knew instinctively that something was going to happen.

And it did.

The moment Lee's foot touched the floor on the other side of the line, the ceiling above him moved, rotating silently like a top as it came down. After a very long second that seemed to drag on and on, the bottom of a ramp extended and stopped right at Dr. Lee's feet.

Now filling the center of the room was a spiral ramp winding its way gently to the next floor.

Hank had no doubt that if Lee had stepped over that line at any point around that circle, that ramp would still have stopped right at his feet.

"Flat amazing," Lee said, shaking his head.

"This place is really starting to get eerie," Edaro said, the golf ball vanished into a pocket.

Hank knew how he felt. It was as if the city was just waiting to do what they wanted, waiting to open up to them. And now that they knew how to get inside, there was no telling what would happen.

No telling at all what this very alien city hid.

8

Time: 11:23 A.M. Pacific Time
9 hours, 52 minutes after Arrival

S tephanie was glad they'd decided not to go up that ramp to that dark second floor. The air coming from above smelled stale and very old. Both Hank and Sergeant Malone suggested there would be no point. They had other buildings to explore and a power source to find, one buried somewhere in the middle of all those skyscrapers ahead.

And who knew how much time they had to find it.

So they had all gone back out into the shadows of the street and continued on their way, the four soldiers moving ahead of the main group, leapfrogging along both walls, investigating any hole in the wall and all side streets.

Nothing moved except them.

The silence felt almost oppressive, as if she were standing at a funeral and no one had anything good to say.

The deeper they got into the city, the more she began to dread what was coming, even though she had no idea what it might be. The street was getting darker and darker as the buildings around them got higher, blocking out more and more of the light from the Maw. She guessed that by regular human standards, the structures lining the street were twenty to twenty-five stories tall, all touching except where the cross streets broke the walls.

As with the shorter buildings, they all looked the same: featureless and white. No windows and all the corner doors closed tight. Also, even though the street hadn't changed in dimension, it felt much, much narrower than when they'd started.

She glanced around at the others after they'd crossed about ten intersections. Everyone else looked as edgy as she felt. They mimicked Sergeant Malone's head movements as they went. Back and forth, back and forth, scanning all the buildings and the street ahead.

Even though they were moving slowly, this was one of the most tiring treks Stephanie had ever taken. And not for one step of it had she taken her hand off the pistol in her pocket.

Suddenly, Sergeant Malone put a hand to the side of her helmet where her earpiece was. Then she looked up. "Halt. Take up standard cover positions."

She signaled that they should follow her over to a spot against one wall. "Waters," she said to one of her men, "I need the orbital link up and running. Hook in all cameras and audio feeds. Stat."

Waters nodded and hopped to it immediately.

"Something wrong?" Hank asked.

Stephanie came up to where she could hear what was happening.

"Got a scrambled message from the facility," Malone said, her gaze scanning the nearby streets as she talked. "They're losing the camera images and audio feeds we're sending back. The buildings here are too tall for regular commlinks, which means we have to reroute everything through our orbital stations."

Stephanie didn't like the sound of that, but she kept her thoughts to herself. Even with the cameras running and communications solidly in place, she doubted that anything was going to make her feel better until they were safely off this ghostly place and back on the mainland.

Private Waters handed Malone the small headphones. She quickly put them on, adjusted the mouthpiece, and said, "Island Mission to base. Repeat previous transmission."

She waited.

Around them the silence seemed to weigh even more heavily as they all waited, too.

Finally, Malone said, "Copy that. Understood."

She took off the headphones and handed them back to Waters. "Pack it up and be ready to move."

Waters nodded and set to work.

"Well," Hank asked, "are the cameras back on-line?"

"They are," Malone said. "But we're going to lose them when we go inside buildings. We don't have the power to boost them through the buildings and all the way to orbit at the same time."

Stephanie looked at Malone, whose usually impassive face suddenly showed some concern. It was the first time Stephanie had seen Malone show even a hint of emotion.

"There's something more, isn't there?" Stephanie asked.

"We've got company," Malone said. She keyed her commlink open, then said to both the scientists and her men, "Listen up, people. What main base thinks is an alien craft just came in, breaking through Union defense screens to land on the northern tip of the island. We're going to need to stay extra alert."

"What?" Bogle almost shouted, his voice echoing off the buildings around them.

"Shit," Stanton said, much more softly.

Stephanie was having trouble grasping what she was hearing. Up until just a few short hours ago, she knew nothing of aliens in the Maelstrom. Now she was standing in an old alien city and some kind of alien aircraft had just landed. This couldn't be happening.

"We need some answers," Hank said, facing Malone. "What do you know about these aliens?"

Malone stared at him without a sign of emotion, then glanced at Stanton. "We're not sure, but they think it might be Pharons. You want to brief these people, Dr. Stanton? You were there."

In the deadly silence of the buildings around them, they all turned to Bradly Stanton. Stephanie forced herself to take a deep breath.

Stanton shrugged, but Stephanie could tell he was frightened.

"Two years ago," he said, "a Union expedition found a small planetoid we now call the Cache because of all the information we found there."

"Were you there?" Hank asked.

Stanton nodded. "As one of a team trying to find out what happened and to secure any technology for study. I've worked on nothing else, basically, for the past two years."

"So the Union has known there were aliens for two years?" Bogle asked. "And told no one."

"By and large," Stanton said. "We know of three different alien races that inhabit the Maelstrom at this time. We call one the Shard, the second Growlers, and the third Pharons."

Stephanie still couldn't really grasp what she was hearing. This was changing everything she had ever assumed about the world around her.

"So which aliens were on the planetoid?" Hank asked.

"Dead ones," Stanton said. "About a year ago my team managed to decode a five-minute recording of part of the battle between the original inhabitants of the Cache and the race we've named Pharons. It was a brutal sight."

"The Pharons wiped them out?" Lee asked.

"Completely, it seems," Stanton said, his hands

moving in that nervous fashion again. "And from what we could tell, without mercy."

"So what do these Pharons look like?" Edaro asked.

"At one time, humanoid in shape," Stanton said, his eyes distant as he remembered. "But from what we saw and the evidence we gathered in other places on the Cache, it seems they are now nothing more than decayed, walking mummies. Most of them are just shambling husks of gray flesh, held together by beautiful and ornate golden armor. They seem to wear some sort of pack on their back that supplies their body with fluids."

"What?" Bogle said. "You're kidding, right?"

"We're standing in a deserted alien city," Stanton said. "I would not kid."

"That description is accurate," Sergeant Malone said. "I have seen the film. But we don't know for sure that the ship that landed is Pharon. We only know an alien ship is here."

Stephanie glanced at the sergeant, who seemed to be back to her cold, normal self.

"We have schematics of the Pharon armor and the life-support tanks that pump fluids through their dead bodies," Stanton went on. "All very advanced technology, far beyond anything humans have come up with."

Stephanie just wanted to sit down. Or better yet, get off this infernal island and back inside the facility with its blast doors. There she might feel a little safer. But she doubted she would ever feel completely safe again. Not with knowing there were aliens all around them.

"The Pharons were called The Rotten by the inhabitants of the planetoid," Stanton said. "We've still got a great deal of data to decode, but from what we've got so far, the Pharons seem to attack without warning and without reason. They take both living and dead prisoners, turning them into slaves to fight and work for them."

"So if they kill us, we could still end up their slaves?" Edaro asked.

"That's right, from what we can tell," Stanton said. "Again, we've only been working on this for two years. We have a long way to go."

Silence again descended until finally Hank turned to Sergeant Malone. "So base doesn't know what kind of ship landed? How big?"

"Small ship," Malone said. "Only one. They do not know what type."

"From the information we got on the Cache, if it is Pharon, it would mean maybe a dozen Pharons inside," Stanton said. "But it doesn't matter how many there are. I doubt we're a match for even five of them."

Malone snorted, a sound that surprised Stephanie. The sergeant turned and glared at Stanton. "I think you either overestimate the Pharons or underestimate my Special Ops squad."

Again Stephanie was stunned. She had no idea that the Union troops with them were Special Ops. No wonder Malone had them working so smoothly together. They were the best the Union had.

This time Stanton snorted. But he didn't argue with Malone.

The last thing Stephanie wanted to do was end up

as a reanimated corpse slave of some mummified alien race. The thought made her stomach twist into a hard knot of fear.

Hank stared at Malone. "Are you recommending we continue?"

"I'm not recommending anything," she said. "This is our mission, and we haven't finished it yet."

Stephanie was impressed that Hank held his ground. He stood there, eye to eye, then after a long few seconds nodded.

"You're right," he said. "We came here for a reason, and we need to keep going."

"What?" Stanton said. "Are you crazy? I've seen what those aliens do to other creatures. Trust me, we don't want to run into them."

"I'm with Stanton, and I haven't even seen the tapes," Bogle said.

Lee nodded agreement. "Yeah, going on sounds somewhat suicidal, don't you think?"

Hank glared at Stanton, then at Bogle. "We came here to investigate," he said, his voice cold and very intense. "And to gather as much information as possible about this place. We have a job to finish."

"Not under attack we don't," Lee said. "At least I don't. I was hired to do research, not fight aliens."

"Look," Hank said, "the aliens are here, which means they think there's something worthwhile on this island. I think we should try to find whatever it is first, at least until Major Lancaster calls us back. And at the moment, they aren't doing that, so they must not feel it's time yet. Am I right, Sergeant?"

"I'll check," Sergeant Malone said. "Island Mission to base."

She waited as Stephanie and the rest watched.

After a moment Malone said, "Confirm status of alien craft and Union response. Is our mission in jeopardy?"

She listened for a moment without so much as a blink of an eye or a nod of the head.

Stephanie could feel her stomach tightening even more, as if the silence around them was getting heavier and heavier. If it got much worse, she wouldn't be able to breathe.

Finally, Malone said, "Understood. Out."

Then she looked directly at Hank. "The alien craft is confirmed to have landed on the northern tip of the island. Union ships have it under surveillance and have monitored no movement at this point. Our forces are building strength for a possible attack. Our mission hasn't changed, with all standby and evacuation measures still in place."

"Good," Hank said. He glanced around at the others. Stephanie had seen a lot of different looks in Hank's eyes over the past few months, but never one this cold and calculating.

After a moment of no one saying anything, Hank turned back to the sergeant. "We're ready to go forward when you are."

She nodded and spoke into her commlink to her men. "Two by two on point. Move it out. Pick up the pace."

Then she turned back to the young private beside her. "Make damn sure you keep one eye on that motion detector. I want a warning if something's coming at us."

"Yes, Sergeant," the private said.

Malone strode into the middle of the street, not waiting to see if the civilians were following.

Stephanie didn't hesitate. With one hand on the warm metal of the Pug pistol in her pocket, she moved into position beside Hank and behind the sergeant.

"We're dumber than we look," Bogle said, falling in.

"No one could look this stupid," Edaro said.

A moment later they were walking at a slightly faster pace toward the towering alien skyscrapers ahead.

And in all her life, Stephanie could never remember being so afraid.

9

Darkness in the middle of the day.

The towering skyscrapers let little of the Maw's white light reach ground level of the city, making it seem as if they were walking the streets in a perpetual twilight. Even when looking up at the sliver of blue sky far above, there didn't seem to be anything bright. Just gloom.

Hank couldn't even imagine what this place was going to be like in the dark. He pulled the zipper on his jacket up closer to his chin. The air was damp and smelled like mold and rot. Everything was so still, as if not even the wind was allowed to blow in these alien canyons.

Hank glanced around at the high, plain walls and side streets. Everything now looked threatening and ominous. The shadows were deeper, making the crumbling holes in the sides of the alien buildings look like black death pits. And out of the corner of his eye he was always seeing something move, but when he turned there was nothing there.

At least at the moment.

He wasn't letting himself think about the alien ship that had landed on the north end of the island. So far they'd heard no sounds of combat or other commotion. He wondered if it was because the sounds weren't audible this deep among the high buildings.

For the last twenty minutes of their advance up the street, no one had spoken. They had simply continued on, their progress unimpeded on the wide avenue. But the moment they got in among the skyscrapers, more and more debris filled the street, until finally the road ahead was completely blocked by the remains of what had been the top fifty stories of a nearby building. It was piled a good three stories high and covered most of a block.

Impassable.

Malone called a halt and ordered a guarded perimeter around them while Private Waters checked the location of the energy source they were using as a Holy Grail.

"Ahead four hundred paces and a few hundred paces to the north," Waters reported.

Toward the north, the direction the alien ship had landed.

There was no way Hank wanted to bump into them, but he knew Malone was immune to protest. He forced the idea of alien soldiers out of his mind and looked over the massive pile of debris up the street. He couldn't see which building the power source might be coming from. More than likely it was blocked from sight by the other structures towering around them. But something must be there.

"We go north then," Malone said.

"We've got movement, Sergeant," Private Waters said, turning the motion detector around so that she could see the face.

"Here we go," Bogle said softly to himself.

"I knew we should have gone back," Stanton muttered. "I just knew it."

Hank ignored Stanton and went to look over Malone's shoulder at the screen.

From what he could see, there were three scattered single beings on the other side of the mound of debris, marked by red dots on the screen, moving slowly along what seemed to be an open street, heading south.

"Could be anything," she said. "More than likely a few animals."

"Or the residents of this place," Stanton said from behind Hank.

"Any chance of Neo-Soviets?" Stephanie asked.

"No," Malone said. "None have landed on the island. I would have been informed."

"How about we check with base to see if those things moving might be the aliens," Stanton asked.

Malone threw him a cold look. "Nothing has yet to leave the alien ship. I've been getting steady reports."

Stanton looked relieved.

"So you're saying," Lee broke in, "that whatever is moving on the other side of that debris pile is something brand-new that we don't know about?"

Malone didn't answer.

Edaro had his golf ball out again and flipped it into the air and caught it, a mute commentary of his own.

Silence filled the street.

"We head north up the side street," Malone said, speaking into her commlink and to the scientists at the same time. "Same point and cover formation for two blocks, then we stop to see if we can get through. Let's move out."

The four squad members who had stopped ahead of them near the wall of debris jogged quickly past, immediately starting their standard leapfrog cover headed north along both walls on both sides of the street. Malone waited until they were a good distance ahead, then led off, moving to the middle of the narrower road.

Hank took his position behind her and glanced over his shoulder as the two soldiers who were guarding their rear fell in, too. He certainly had to hand it to Malone. She ran a very tight, very well disciplined squad. And he was damn glad they were Special Ops.

The rest of the group looked serious and obviously unhappy with the situation. Bogle's face looked pale, and even Lee wasn't smiling. Stanton looked like a walking ghost, with sweat dripping off his face. More than likely it was because he could more easily imagine what kind of horror they might be facing. It was enough to scare Hank half to death just hearing that aliens had landed. He couldn't even imagine his terror if he knew anything more.

Stephanie seemed to be doing just fine, but her hand never left the pocket with the pistol. Hank, too, was very glad he had the rifle in his hands. He didn't want to have to use it, but it was good to have it.

A lot of debris from the buildings above had fallen onto this street also, but they managed to pick their way over it for the first block. At the next intersection they saw that the street leading west was also blocked.

"We still have movement, Sergeant," Waters said. "Pacing us two blocks to the west."

"How many?" Malone asked.

"Still just three," Waters said. "No cover formation, just slow movement up the middle of that street there."

Malone only nodded and didn't miss a step.

Hank didn't bother to ask why she thought something or someone would pace them. She wouldn't have any more idea than he did.

At the next intersection they stopped and took cover against the side of one structure. The way was also blocked to the west. The pile of fallen debris was even bigger there, filling the street to a height of at least four stories. There was no going over it.

Without being ordered to do so, Private Waters quickly set up the equipment to detect the energy source. "We've gone too far north," he said. He pointed at the pile of debris. "Over that about twelve hundred feet. And down some."

Hank glanced in that direction, but still couldn't pinpoint which building the energy source might be in.

"The three still with us?" Malone asked.

"Stopped on the other side of that pile," Waters said.

Again, Malone only nodded.

"Feels like a guarded fort, doesn't it?" Hank asked, staring at the massive wall of debris.

"It does," Malone said.

"So someone, or something, doesn't want us in there," Bogle said. "Piled up tons of debris to keep folks like us out."

"Seems that way," Hank said.

"But why?" Stephanie asked.

"Maybe because they live there," Stanton said. "Why else build a fort?"

"To protect something," Lee said.

"That too," Stanton said.

Malone was about to speak, then stopped and glanced around.

It took Hank a moment to realize what was happening as the ground under him began to shake slightly. Then the trembling grew in intensity until he had no doubt what it was.

Earthquake.

A good-sized one.

And they were standing between crumbling old ruins of very tall buildings. The worst possible place to be.

But inside the buildings would be even worse.

"Center of the intersection!" Hank shouted.

He grabbed Stephanie's arm and they tried staggering that way as the ground moved beneath their feet and pebbles started to rain down around them. Hank only hoped pebbles were the worst that would hit them.

The roar seemed to build, as if the entire city was angry. Dust filled the air, and the sounds of hunks of buildings crashing to the ground surrounded them like thunder.

Hank stumbled and fell, but quickly scrambled up with Stephanie's help, his rifle tight in his hand.

"Center of the street! Everyone!" Malone ordered her troops. "Move it! Move it! Move it!" She grabbed the energy-sensing equipment and sheltered it against her stomach as she moved surefootedly away from the wall.

Bogle grabbed Stanton's arm and dragged him away from the building as the others joined Stephanie and Hank.

Then, as quickly as it had begun, it was over.

Hank couldn't believe it. He and Stephanie hadn't even reached the center of the intersection.

White dust billowed in the still air like slow-moving clouds in the sky.

Hunks of a building smashed down just to the east of them. Small, pebble-sized pieces of the buildings continued to rain down on them.

But nothing big.

Hank wanted to look up, but didn't dare. Instead, he kept one arm over his head and waited.

Slowly everything settled, and the falling pebbles stopped.

As best as Hank could see in the dust, no one was hurt.

Bogle and Edaro started coughing, as if they were choking. Lee patted Bogle on the back and he finally stopped.

"Return to your cover posts," Malone ordered her men. "And report in. Injuries?"

She listened for a moment as Hank watched, trying to take shallow breaths of the dust so he wouldn't choke.

Finally, the sergeant said, "Good. Stay alert, people."

Waters was already on one knee checking the motion detector.

"Three still with us to the west," he said. "Nothing else moving."

"Except the ground," Bogle said.

Malone glanced around the group, checking to see if everyone was all right. Hank did the same.

Bogle, who was tall and normally very straight-backed, was slightly stooped from the coughing, but slowly seemed to be getting his breath back.

Lee now looked much older, his golden brown hair turned nearly white by the dust. And he wasn't smiling at the moment, which added to the effect.

Edaro seemed fine, the golf ball still in his hand as he worked at something in his pack.

Stanton's eyes were wide in his dirt-covered face, the fear obvious in his every movement.

Stephanie looked frightened, too, but seemed to be hanging in there. The dust had covered her hair, too.

The dust slowly settled around them, and the air gradually cleared. Hank kept breathing slowly, not wanting to take too much of the alien material into his lungs.

Malone gave base a quick rundown of what had just happened, then turned to look at the pile of debris.

"All your men all right?" Hank asked as he brushed off the arms of his jacket.

Malone nodded. "No injuries."

"We were lucky," Stephanie said.

"Boy, you aren't kidding." Bogle spat out more dust, then began to cough again.

Lee simply dusted off his jacket and said nothing.

Hank had to agree with Stephanie, looking at all the massive hunks of alien rock that had fallen in the past. They were damn lucky.

"We've got more problems," Edaro said, but broke off with another fit of coughing.

Hank turned to see him kneeling in the middle of the street, studying an instrument he was trying to keep clean of dust by sheltering it with his body and a back pack. His trademark golf ball was gone for the moment.

Jeff Edaro was a specialist in molecular science, which had made him a natural to investigate the molecular structure of this island and the stuff on it. That made sense to Hank. Anything that could just appear somewhere had to be doing weird things on a molecular level.

The instrument Edaro knelt over was the size of a

laptop computer, with a dozen silver prongs pointing off the back of it. It was making a slight beeping sound as it worked. On the screen it seemed to be scrolling graphs of different types and configurations.

Hank, Bogle, and Malone went over to see what Edaro was up to.

"What are you talking about?" Bogle asked.

"That wasn't just a standard earthquake," Edaro said.

"Felt like one," Bogle said. "I'd guess about a six-point-five on the scale."

Edaro shook his head. "Sure it was an earthquake, but it was caused by a molecular phase—a time-space event that covered the size of this entire island."

Hank had an idea what molecular phasing meant, and he didn't like the way Edaro used the term time-space at all. He didn't know that anyone studied it outside of theoretical physics, but at the moment there was nothing theoretical about their situation.

Edaro glanced up at Hank, then over at Sergeant Malone. Then he took a shallow breath, coughed once, and went on. "We all know this island didn't just travel here. It appeared here."

Hank nodded.

"So it didn't happen with fairy dust," Edaro said. "And no one hit it here with a five iron. Something objective had to have occurred. Since the middle 1990s, there have been experiments in molecular phasing."

"Transporting one object instantly from one location to another?" Stephanie asked.

Edaro nodded. "The first success was in Australia in 1996, with only a single atom. My guess is that this entire island, every molecule of it, including the buildings, sort of phases out of this space and transports to some other time and place. Maybe the way Earth got taken into the Maelstrom. Only we don't know that for sure by any means."

"How do you know all of this?" Malone asked, staring at Edaro.

"It's my field of study, Sergeant," Edaro said, laughing, his teeth looking extra white against the dirt on his face. "I've been working on something similar on a much smaller scale for the Union for the past five years, but so far without much success. Actually, a lot of us are working this area, both as weapons research and as a possible way to escape the Maelstrom and return to normal space."

"You're saying this island almost left with us aboard?" Bogle asked.

"No, not that time," Edaro said. "Not a strong enough molecular phase to shift everything. But close, like a car engine turning over, but not starting, I'd say."

"Not this time," Hank said. "But maybe next time we go for a ride? Is that what you're saying?"

Edaro nodded. "Exactly what I'm saying. Right now, on a molecular level, this place is as stable as anywhere else on Earth. But I'd bet anything it won't remain that way."

"Wonderful," Bogle said. "What a great place for a vacation. Aliens on one end, who knows what the hell is lurking over that debris, and now the entire island might leave with us aboard. I've got to talk to my travel agent when we get back."

Lee laughed. "Part of the package, remember?"

Bogle nodded. "Oh, I remember. I've just changed my mind is all."

Hank ignored Bogle's attempt at humor and asked, "Any way of telling when the next phase event will occur?"

Edaro shook his head. "None that I'm aware of. It's been almost eleven hours since the island appeared here. My guess, and it would only be a guess, is that the next event will happen before the next eleven hours have elapsed. Maybe considerably less."

"Why do you say that, Doctor?" Sergeant Malone asked.

"Simple," Bogle said. "From the looks of what we've seen so far—the watermarks on the walls, the debris—this city has been jumping around the Maelstrom for a long, long time."

Lee and Edaro both nodded. Hank had come to the same conclusion.

"Which means the aliens have been tracking it," Stanton said.

"What?" Hank asked. Stanton's statement made no sense at all.

"The aliens in that ship, whoever they are, almost beat us here, and we were only a few kilometers away. So somehow they must have known the location of the island's next jump. Or had a way of tracking the jump when it happened. That would explain why the first place on Earth they landed was this island. There must be something here that they want pretty badly."

"Possible," Edaro said. "But way, way beyond what we've been working on over the last few years.

Though I suppose it would still be theoretically possible to track a molecular phase."

Hank watched as Edaro seemed to go off into his own thoughts for a moment.

Malone knelt down beside where Edaro was staring at the strange-looking instrument. "Can you tell what caused the last event?"

Edaro laughed. "Not even close. But there's something about this island that causes the phasing to another location. My guess, again just a guess, is that the pressure of some sort of 'energy' builds until finally the entire city shifts to a new location."

"Like a fault line building up pressure, then releasing?" Stephanie asked.

"Or a battery gathering a charge," Bogle said.

"Either one would be a good way of looking at it," Edaro said. "But please understand, I'm really only speculating based on my research over the past few years. One thing I can tell you for sure—that last earthquake was no fluke. The island tried to phase out of here, but there wasn't enough energy for it to happen."

"Yet," Bogle said.

"Yet," Edaro agreed.

Hank didn't like the sound of that word any more than the rest of them did.

10

Time: 12:40 P.M. Pacific Time
11 hours, 09 minutes after Arrival

T he dust that had choked the air had just finished settling when Waters said, "They're coming."

Those were not the words Malone wanted to hear, but they were the ones she'd been expecting. Because they had stopped for so long, it gave whoever or whatever was on the other side of that debris barrier in the street time to decide to come after them.

"How many?"

"Just the three," Waters said.

"Move back," she ordered into her commlink.

"Form up. Defensive positions around this intersection."

"What's coming?" Downer asked. He and Bogle were standing beside her.

She held up her hand for them to wait a moment, then asked Waters, "Where are they coming through?"

"Two through the building to our left," Waters said. "One from the right. And now there's a fourth, a new one, coming in from the south along the same street we came down."

That meant they weren't surrounded. A good sign at this point, especially in dealing with an unknown enemy.

Malone turned and pointed to a large pile of rubble near a far wall. There was a small hole in the building there, a defensible position if needed.

"Vasquez. Marva. Get the civilians inside that hole." She pointed, and Waters nodded.

"Clear the interior first," she ordered. "On the double."

She glanced down at the private watching the motion sensor. "Waters, take cover with the civilians and stay on that. I want a running report on their movement. I need you to be our eyes."

"They're going to be coming through soon, Sergeant," Waters said, grabbing the motion detector and following Vasquez at a run.

Malone turned to Downer. "Get your people into that hole and fast. Be ready to fight."

Downer nodded and without a word shoved Dr. Peters and Stanton at a run across the street toward their cover. Dr. Edaro had his equipment under his arm and was also running just ahead of Dr. Bogle and Dr. Lee.

"Okay, people," she said into her commlink, "I want to meet them head-on. Fire from cover and make each shot count on my orders. Do not, I repeat, do not let them get close to you for any reason. Fall back if you have to."

She had no idea what they were facing, but staying away from them was the best strategy under the circumstances.

Malone eased down behind a large chunk of stone that had fallen near the corner of the intersection. Out of the corner of her eye she could see Downer making sure the last of his people were inside, then duck in behind them himself. A moment later Vasquez took up position in the debris in front of the hole, Pitbull rifle out and at the ready.

Three of her men inside, five out on the street, plus her. Six against the four coming, with Vasquez from the hole if needed. Good odds, depending on the fighting ability of whatever, or whoever, was coming.

It looked as if she was just about to get some firsthand information about aliens.

"Two moving toward an opening in the building on the right, near the debris wall," Waters said over the commlink. "One coming into the open now from the left."

"Easy now," she said. "I want clear shots. Wait until I give the order. Clean and simple."

From the building on the left a figure appeared. In the remains of the floating white dust and the gloom of the street, she couldn't see it well. But it seemed tall, with a flowing black robe that swirled around it much as the dust swirled in the street between them.

It didn't seem to be carrying a weapon, but that didn't mean one wasn't hidden in the black robe.

Two more of them appeared to the right, coming through the hole in the wall and seeming to float down over the piled debris, as if they had no feet. For all she knew, they didn't. Only robes and hoods over their heads, hiding their faces in dark shadows.

Maybe they didn't even have faces. Tall, thin, and flowing-robed humanoid shapes. No other details were visible.

Weird.

Just flat weird. In all her career, she'd never had to fight an enemy completely unknown to her. She didn't like this one bit.

As the creature came into the open it was as if the temperature of the street dropped twenty degrees. She could feel the chill coming off of them.

Cold.

Deadly cold, she would wager.

The alien city already had a cold feel about it, but now it had gone completely arctic.

"Two more coming in from the right down the street," Waters said in her earpiece. "Another from the left, also on the street."

That made seven total creatures. The odds were evening up a little.

The Bulldog felt good in her hands as she laid it across the top of the rock and took aim on the robed figure on the left. She couldn't see a face in the shadow of the hood. But she wasn't going for a head shot anyway.

Around her the air temperature kept dropping, as if the creature was sucking all the heat out of it. Suddenly her gun felt like ice in her hands.

The three creatures in the open moved purposefully, flowing like water directly at the three men nearest them. It was as if they had no fear.

Or any idea what they were walking into.

The cold seemed to have become intense, like a weapon. She could feel it biting at her skin, numbing her joints, draining the life right out of her.

"Fire," she ordered, her voice almost choked off by the sudden strange cold. Her breath froze in front of her.

The canyons of the alien city instantly filled with the roar of gunfire as they all opened up on the three creatures moving out in the open.

Her rifle kicked in its familiar way in her hand as she sent the creature she was aiming at tumbling backward into a pile of black robes. But the cold made the kick painful. Had she waited another few seconds, she might not have been able to squeeze the trigger.

Too close.

She caught a glimpse of blackened skin under the robe of one creature, but nothing else to tell her what they really looked like.

All three creatures were down.

The firing stopped as quickly as it started.

And around them the air seemed to warm almost instantly.

But the creatures weren't staying still.

Malone watched the one she had shot as it simply dissolved into a pile, the robe seeming to vanish until there was nothing left but what looked like a mound of very black sand.

Sand as black as their robes.

Then, above the pile the wind started to swirl around and around, a small twister coming out of the center of the pile, picking up a few grains of the black sand at first, then slowly getting bigger and bigger, drawing in more and more of the black sand as it gained power.

Within seconds the area of the intersection was filled with three large tornadoes, growing in force, swirling not only the sand into the air, but clouds of white dust and debris with it.

The cold was still intense, as if the wind was blowing off an iceberg.

Malone could feel the force of the winds pulling at her, at the rifle in her hands. It was as if the sand things were trying to disarm her after they were dead.

She held on, clutching the rifle tight against her chest with her numb hands as the winds gained in intensity and strength, wiping at her with icy fingers.

"Stay down!" she ordered her men.

In the hole she could see Vasquez and Dr. Downer duck back inside.

The winds whipped around her, forcing her to snap her helmet's visor down into position to protect her eyes. She watched as the three tornadoes joined into one large swirling black wind rising a good ten stories into the air.

Then, as quickly as they grew, the winds seemed

to form together into a black stream, sucked over the barrier of debris in the direction of the power source.

And again there was no wind.

Just calm, very cold air.

And a very quiet dead city.

What the hell had just happened?

Where there should have been three alien bodies there was nothing.

Her mind balked at what it had just seen. They had been too easy to kill, yet had they killed them?

And how close had they come to being frozen by these creatures? Her fingers were still numb.

"The other four are still coming," Waters said over the commlink.

"Hold on, people," she said. "Only fire when we've got a couple of them in the open."

"I can see two of them from here, Sergeant," Hawk said. He was in a position behind some debris on the corner, looking south. She couldn't see down that street from her position. "And it's getting damn cold again."

"Copy that," Cort reported, his voice weak-sounding through the commlink.

Cort was in a similar position just around the corner to the south from her. But he would be closer to the alien creatures and thus more exposed to the cold. Much longer and he wouldn't be able to fire. She didn't want to be losing men to the cold, that was for sure.

"Got one from the north," Harden said. "Damn cold."

"Copy Harden," Raynor said.

"Waters, location of the fourth one?"

"Stopped inside the building where the first two came out," Waters said. "Looks like it's staying there for the moment."

Again she could see her breath in front of her. It had to be even colder closer to those things.

"Take out the ones you can see, people," she ordered. "Make it clean."

The burst of gunfire again filled the canyon between the buildings with a roaring sound that was quickly cut off. The echo drifted down the streets and then died away.

"Two down to the south," Hawk said.

"One down to the north," Harden said.

"Hang on, men," she said as again the wind started to swirl strongly on both sides of the intersection. "Protect your equipment. Stay out of the black sand."

To the north she could see a large tornado climbing into the air between the buildings, the black sand swirling in it.

Then it moved out over the intersection above her, joining with the two from the south, again whipping at her uniform and helmet.

Once more she felt as if the wind, with its cold intensity, wanted to yank the rifle right out of her hands. She held on tight as the black winds spun together, climbing higher and higher until finally they had completely joined into one large, black swirling mass.

Then, just as the first three had done, the wind and sand flashed off over the debris barrier in the street in the direction of the power source.

The silence that again filled the intersection was charged. The air was crisp and cold.

But much warmer than it had been just moments before. And getting warmer by the second.

"The last one has moved off," Waters said. "Clear."

"Report," she ordered. "Casualties?"

One by one her men reported in. No injuries. Just cold hands and fingers.

"Vasquez, bring out the civilians."

She moved over to the spot where the creature she had shot had fallen. There wasn't any sign of it at all.

She studied the ground around the spot. The street had been completely scoured of any trace by the wind. It was as if the creatures never even existed. Yet those nonexistent creatures had almost killed them by sucking all the heat from their bodies. A pretty powerful weapon.

She turned around to face the angry face of Dr. Downer.

"All right, Sergeant," he said, his voice amazingly calm considering the look on his face, "I'm assuming there are a few more things about this island we weren't told. Such as those creatures."

She shook her head, again glancing down at the place where the black-robed being had fallen. "Didn't know a thing about them. And I have no idea what the hell just happened to them, either."

"Nothing?" Downer asked.

"Not a damn thing," she said.

"What do we call them?" Bogle asked, moving up beside Dr. Downer.

Malone only shrugged. It wasn't her business to name alien creatures. Her job was to defend the Union and these scientists against them.

"Sand would be as good a name as any," Downer said. "At least from what we just saw."

"That, Doctor, is you and your colleagues' department," Malone said, turning to rejoin the rest of her men. "My job is just to kill them if they threaten us."

And she had a sneaking hunch that over the next few hours, she was going to be doing a lot of that.

11

Time: 1:03 P.M. Pacific Time
11 hours, 32 minutes after Arrival

Stephanie could barely grasp what Hank and Sergeant Malone had explained to her and the others who hadn't seen what happened in the fight. These creatures, when shot, dissolved into a pile of black sand and blew away. It wasn't logical, it didn't make sense how it could happen or even that it had happened. Nothing in her medical training could explain that such a physical event was even possible.

Yet it seemed, from the witnesses, that it was. But even with that, she couldn't believe it. She was about to ask Malone another question when Edaro pointed to the molecular sensor he'd being using to study the earthquake before the attack. "Would you look at this!"

"What?" Bogle asked, moving over to stand above Edaro. "We going to have another earthquake?"

Edaro ignored the question and scanned back through the last twenty minutes of the instrument recordings. "Here's the earthquake." He pointed to a large spike on what looked to Stephanie to be a straight-line graph. "There was a molecular disturbance in everything on the island."

"You told us this before," Bogle said.

Edaro held up his hand for them to wait as he moved the screen graph along to a different spot. He pointed to two other smaller spikes. "Here and here are when the aliens were killed."

"You're saying," Stephanie said, "that the aliens phased out of this space, just as the island did when it left wherever it was and jumped itself here?"

"That's exactly what I think happened," he said.

"Are they dead?" Malone asked. She'd been standing to one side, listening intently.

Dr. Edaro glanced up at her and shook his head. "If I were to make a wild guess, I would say no. Your shots caused them to phase, and their parts, the black sand that you described, were drawn to the energy source."

"Where they were re-formed?" Hank asked, shaking his head. "I don't think I can believe that much."

Stephanie couldn't either. Her mind still boggled at the idea of an alien body dissolving into sand and creating enough energy to cause the winds Malone described.

Edaro only shrugged. "I'm just making a wild guess, but from what the sergeant tells us and what my instruments show, it wouldn't surprise me that with the energy source, they just might. This entire island basically dissolves and re-forms on a molecular level every time it moves."

"Centered around that energy source we're picking up?" Malone asked.

Edaro shrugged, his golf ball back in his left hand. "Again, your guess is as good as mine on that one. But *something* has to be powering this, or maybe attracting the energy of the Maw in some alien fashion. To make these shifts takes a lot of energy. That energy has to be coming from somewhere; otherwise, this city wouldn't continue moving around in space like it does."

"And the most likely suspect is the energy source we're reading," Hank said. "Right?"

"Possibly," Edaro said, flipping the ball into the air and catching it. "It's the only one we have so far, I assume." He glanced at Sergeant Malone.

"It is," she said.

"Well," Edaro said, "if you need an energy source and you only have one on hand, it seems logical to look at that one first."

Stephanie took a step back and tried to get a deep breath of the stale and cold air. By now, she had managed to grasp, or at least accept the idea of an island simply appearing off the Oregon coast. The Maelstrom had vast power that often followed no natural law of physics.

But these were alien beings they were talking about. Beings who'd have had to evolve, have had some form of continuing the species in some material fashion, alien or humanoid. And that was what she was having problems understanding.

Or maybe she was making wrong assumptions. As Stanton kept reminding them, applying human assumptions to alien phenomena could very well be wrong. Dead wrong.

She turned to Hank and Chop Edaro. "Is it possible that the dissolving and possible re-forming of these creatures wasn't a natural part of their evolution?"

"Your guess is as good as mine," Edaro said. "But I would say that's more than likely. Just as shifting to different parts of the Maelstrom wasn't the natural state for this city when it was built."

"Or being in the Maelstrom isn't a natural state for Earth," Lee said, shaking his head.

"Exactly," Edaro said.

"So these Sand creatures may have become the way they are because of the shifts," Hank said. "That would make sense."

"And if we're here when the island shifts," Stanton said, joining the group, "do we become like them?"

Stephanie felt like slapping him for saying that. None of them had any real answers. They were using human logic to take one small step at a time, but everyone knew it was like shooting in the dark. And at the moment they didn't need the negative Stanton reminding them of every possible thing that might go wrong.

She glanced at Hank, who was glaring at Stanton. Hank obviously felt the same way about Stanton.

Edaro broke the silence. "Anything is possible. The answers, I would guess, if there are any, are at that energy source."

"You can't be thinking about continuing, after this," Stanton said, glancing around at Sergeant Malone.

"We will move out shortly," Malone said.

Stephanie watched as Stanton's face grew even whiter than it already was. He took a deep breath as though trying to collect himself, but said no more.

Stephanie understood exactly how he felt, but there was no turning back now. She stared up at the towering buildings overhead, the sky bridges, the slivers of sky and light on the buildings far above. This must have been a great city in its time. What terrible accident or quirk of Maelstrom fate far in the past had started it on this jumping road? She doubted anyone would ever know.

"Okay, people," Sergeant Malone said, "let's get moving, see if we can find a way around this debris blockade."

At that moment, Stephanie was staring at a sky bridge and an idea hit her. Instead of looking for a way to go in through the debris, why couldn't they just go up over that sky bridge and down? Just as the original residents might have done.

"Hang on, Sergeant," Hank said, speaking ahead of Stephanie by a half a second. "What about going through where the Sand did?"

Malone shook her head. "Too dangerous. Too tight. Could be booby-trapped And if we get another shaker in that debris, we'd never get out."

Stephanie agreed with that and it made her like her own idea even more. "How about we go over?"

"What?" Bogle asked.

She pointed up. Everyone glanced up, then Malone looked back at Stephanie. "What did you have in mind?"

"The sky bridges," Stephanie said. "We go up here, over the sky bridge, and then down. We bypass the debris blockade completely."

Hank nodded, staring up at the bridges over them. "Just might work."

"And it would give us a chance to explore a couple of these buildings along the way," Bogle said.

Stephanie noticed that Malone was also nodding, probably calculating what might be the military and defense problems of being inside. Finally, the sergeant said, "Worth a try."

She turned and strode back to the middle of the intersection, her boots making no noise on the hard surface. Then she stopped and stared upward again, clearly picking a path of bridges.

Stephanie and the others walked over to join her.

Malone glanced at Hank, then pointed upward. "We go up about fifty flights to that bridge there."

"Yeah," Hank said. "Take it over to the next building."

"And then up another twenty flights or so to the next bridge," Malone said. "Two more buildings on the bridges at that level and we should be just about over the location of the power source."

"Sounds good," Hank said.

"Sounds like a climb," Bogle said, staring upward.

Of course it does, thought Stephanie. No one could imagine that climbing seventy flights would be fun. But so far nothing about this expedition had turned out to be fun. Nothing at all.

And she didn't expect that to change any time soon.

12

The ground-floor chamber of the building was dark, damp, and covered with a thick layer of dust that seemed as light as air and swirled around Hank's feet in eddies like water.

"Pick up your feet and walk lightly," Malone ordered, "unless you want to choke on the dust."

Malone had sent advance men ahead, but they were nowhere to be seen as Hank entered the big, dark room. The central ramp was down, and in the beams of everyone's lights, he could see footprints heading up the ramp. Her men had obviously gone ahead, scouting.

Before entering the building, Malone had contacted base control and reported their plan, knowing that the video and audio uplinks to the orbital stations would be cut off while the group was inside the buildings. Hank didn't much like the idea of being out of touch, but base control gave the go-ahead so he kept quiet.

"We're going up spread out," Malone said. "One floor between us. Two of my men with every two civilians. You hear shooting above or below you, or anything else, you scramble for the nearest wall and find what cover you can. Understand?"

Hank nodded right along with the others. He just hoped her plan would keep them all from getting killed at the same time in an emergency.

"Dr. Downer, Dr. Peters, you two are with me," Malone said. "Waters, take point. Hawk, stay behind us. Let's move. The rest wait for my signal to start up."

The second floor was as dark and as empty as the first. But the dust had decreased to a thin layer. The ramps to the floor above seemed almost to meet at the ramp coming up from below. So at each floor they were only on a flat spot for about three steps, then they started back up again.

The ramps weren't steep, but it took them two times around the wide circles to reach the next floor. By the time they passed the dark third floor, Hank was beginning to wonder if the aliens who'd built the place had thought of elevators. And that maybe he should stop and go looking for them.

Sergeant Malone started Bogle and Stanton and two of her men when Hank and Stephanie were halfway up the ramp to the third floor. And when they got halfway up the ramp to the fourth floor, she started Edaro and Lee and two more men.

By the fifth empty and dark floor, Hank was beginning to wonder what exactly had happened to this city. Someone or something had obviously cleaned it out at some point in the past. They saw not a stick of furniture, no sign of any occupation. But had it been cleaned out before or after the shifting started?

The second, third, and fourth floors were all dark and smelled of mold and dampness. From the watermarks on the walls and the caked dust, these floors must have been flooded at times during the city's hops around the Maelstrom. Maybe the water had simply flushed away every trace of life. Given enough years in enough water, that was possible.

Or maybe the furniture and bodies of the original residents were the dust that covered everything. Given enough years, that, too, was very possible.

"Understood," Malone said in response to something said to her over the commlink as the five of them reached the fifth floor.

"Take a look at that," Stephanie said, pointing upward.

Hank glanced up at the light coming down the opening from above. There must be a hole in the wall up there a few stories. A big hole to allow that much light.

"Seventh floor has windows," Malone said matter-of-factly.

Then, after a short moment's pause, she said, "So does the next floor above that. And no more ramps."

"What?" Hank asked.

She held up her hand. "We'll see when we get there."

Five long minutes later they had climbed into the light of the seventh floor. **It didn't just have windows, it didn't seem to have walls.**

Not one wall or support beam in the entire city-block-sized room.

It was as if the massive floor above was just floating overhead.

Hank was shocked. What had looked like solid wall from the outside was perfectly transparent from the inside.

Frighteningly so.

The weight of the ceiling over their heads seemed to want to come crashing down at any moment.

"This is incomprehensible," Stephanie said, staring around the massive room.

"One fantastic piece of engineering," Hank said. "And I want samples of that transparent wall material before we leave this island."

"We all need to start collecting samples of things we find."

"Good idea," Hank said, reaching down and picking up what looked like a small piece of wall material. He studied the grayness of it for a moment, wondering why it wasn't transparent when not part of the wall, then he slipped it into his backpack.

"One-way transparent walls," Stephanie said, "strong enough to hold up a skyscraper. Amazing."

Hank couldn't have agreed more.

Then he noticed something else. There was debris

scattered in piles around the gigantic, block-square room. Not much, but some. It looked like old floor covering and maybe pieces of ruined furniture. Some of it had piled up against one transparent wall, looking as if it should simply fall out. They were going to be able to get quite a few specimens from this place.

"One more floor," Malone said, not pausing, but continuing to move surely up the ramp.

As they climbed Hank couldn't get over his astonishment at the transparent walls and the view beyond. The building next door was just as solid-looking from high up as it was from below. But he wondered if those walls were as transparent as the ones he was looking through. More than likely they were.

When they reached the eighth floor, it was exactly the same as the one below it. Transparent walls, lots of debris. The only difference he could see right off was that there was no ramp leading upward, just a flat ceiling overhead. Maybe the ramp was just broken. That would be logical, considering what the city had been through.

But it was also logical that its builders would have another way of reaching the top floors besides climbing ramps. After just eight floors, Hank was breathing hard. He didn't even want to think about climbing another forty-two floors just to get to the first sky bridge. And there had to be another hundred floors above that in this building alone.

But there was no other way up. He could see every meter of the huge room and beyond that to all the buildings around them.

He forced himself not to stare out through the

walls for a second and to look around instead. This level contained more debris than the others they'd seen. Piles of broken furniture and what looked like equipment were scattered everywhere.

Hank wanted to search through them, but first he had to check out the amazing walls. He wound his way to one edge of the giant room alongside Stephanie. Eight stories below, the street stretched in several directions. It seemed as if the floor ended abruptly and that he could have just jumped out into space if he wanted to. But if he looked closely, he could see the surface of the wall. It was like the most highly cleaned piece of glass he had ever seen. Yet he knew it had to be extremely thick. And it had to be very strong to form the walls of a skyscraper like this one.

At the same moment both he and Stephanie reached slowly out and touched the wall. It was cold and smooth to his touch. And very solid. Hank licked one finger and tried to smudge the wall, but left no mark at all.

"Not sure I could get used to this," Stephanie said.

"Murder on someone with a fear of heights," Hank said.

Stephanie pointed up at the roof. "I meant the feeling that the roof is going to come crashing down on us at any time."

"Thanks," Hank said, smiling ruefully. "I didn't really want to think about that possibility."

She laughed, and they eased as close as they dared to the thick, transparent wall and looked down.

What seemed like directly below him, but was actually up one block, was the debris barricade filling the street. And he could see all the way down the street beyond the barricade toward the power source.

"Sergeant," Hank called, "I think you might want to take a look at this."

He moved over to the corner of the building for a better look at the barricade below.

Malone came up alongside him and looked where he pointed.

Farther down the street, on the other side of the building that they thought was the location of the power source was another debris barricade blocking the street. He and Malone had been right. Someone, or some people, or some aliens, had gone to a vast amount of trouble to build a blockade around the central area of the city.

"Why would anyone do that?" Stephanie asked.

Neither he nor Sergeant Malone had an answer for her.

They weren't going to find the answer to that question from up there. Somehow they had to get inside the blockade.

"Okay," Bogle said, moving over and joining them. "How do we get any higher?"

Hank looked at Stephanie, and she just shrugged. "I assume the ramp from above just isn't working," Hank said. "We'll have to go back down and try another building."

"I don't think so," Lee said.

Hank turned to face him. Lee was smiling, and beside him Stanton had a bit more color in his face and was looking a little more confident again. Maybe he'd finally managed to surmount his fear. Hank hoped so, for all their sakes.

"Explain," Malone said.

"This is a high-rise," Lee said, continuing to smile. "Granted, a very advanced one with amazing walls and engineering, but still a high-rise building. There has to be an entire core area we're not seeing. Air circulation, supply elevators, heating or cooling, water, and so on. If there are elevators or another way up, we would find it in that core or cores."

Hank agreed completely with Dr. Lee, but as he turned slowly and looked around the monster room, he just couldn't imagine where such a core might be located. He could see into space on all four sides of this building. It felt as if the floor above was floating five meters in the air above them. Everything else on the entire floor was visible.

There was simply no place to put elevators, stairs, and so on.

"So, where would it be?" Malone asked, also staring back at the vast space of the room and the open air beyond.

Lee shrugged. "It has to be here somewhere, I'd bet on it. Maybe inside these transparent walls."

"This is just plain crazy," Stanton said.

Everyone ignored him.

"Okay, look," Lee said. "The builders of this city were masters at certain things, one of which was making materials appear completely transparent from one side while solid from the other. Why wouldn't they be able to move light around an area like a maintenance core just as effectively?"

"Like a magician's trick?" Hank asked, instantly understanding where Lee was coming from. "So it couldn't be seen."

"Exactly," Lee said.

"So we fan out around this room until we bump our noses on something," Bogle said. "Shouldn't take that long."

Malone instantly took charge. "Everyone move to the right wall. Hawk, Cort, Raynor, help us. Waters, stay on post at the top of the ramp and keep an eye on that motion sensor. The rest of you hold positions."

Hank figured she had positioned the others floors below as guards.

They all did as she ordered and lined up along the right wall. But even with ten of them, they were still a good four meters apart along the transparent wall when they started walking slowly toward the opposite wall.

Hank was the second one from the left of their stretched-out, ten-person search line. Stephanie was closer to the edge.

It wasn't until they had worked their way almost back to where the ramp went down that they found anything. He was only a step away from the invisible wall when he saw it. If he'd been moving any faster, it would have hurt.

"Got it," he said, reaching forward and touching the hard surface he was having trouble seeing even as he was touching it.

To his left, Stephanie said, "Here too," as she reached out and touched the invisible wall extending out into the middle of the big space.

"Over here, too," Bogle said.

Directly across from Hank, on the other side of the circle that was the ramp down to the floor below, Bogle and Lee were both running their hands along an invisible wall.

"That looks really strange, doesn't it?" Stephanie said from beside Hank.

Hank laughed. It did look strange, as if the two of them were doing a pantomime act.

"Two cores," Lee said. "Makes sense. Any building this size would need two cores to deliver everything to the floors. And they're positioned to supply the entire space efficiently. These people did just about everything right."

"Except survive," Stephanie said only loud enough for Hank to hear.

"Got that right," Hank said, just as softly.

They watched as Lee moved around the ramp opening, staring at the floor until he finally said, "Got it." His smile was even bigger than normal.

Hank couldn't believe they had missed it again. Lines on the floor. Of course. With the transparent walls distracting him when he came up, it hadn't occurred to him even to look.

Lee turned and walked directly away from the ramp circle until suddenly a door slid open in the middle of what looked like thin air. It was as if the view of the buildings beyond just sort of slid back, showing a white-walled space.

"I'm starting to really like whoever built this place," Bogle said, staring through the open door, then peering around behind it at the seemingly empty space and the buildings beyond.

"Talk about a hidden room," Lee said. "Wow."

Hank was having trouble believing what he was seeing, also. His eyes told him there was no room there, yet when he looked through the open door hanging in the middle of the room, there *was* a room there.

A solid, white-walled, fairly large room.

"Elevator through here, I'll bet," Stanton said, smiling at Malone.

For a brief moment she actually shook her head in amazement, then went right back to work.

"I'd prefer stairs," she said.

She went quickly over to where Stanton was standing and looked inside the door without actually going through it. Hank noticed that she kept her rifle ready.

Then she stepped through with one foot and pressed her back against the door. She glanced at Waters at the top of the ramp. "Anything moving around us?"

"Nothing."

She nodded. "Hawk, go right. Cort, go left. Doors on both corners. Need lights."

Hank moved over so he could see inside the door as Privates Hawk and Cort went in, guns ready. The space was a twenty-by-twenty room with plain white walls. Hank could see clear pull handles in both corners. You could not see through the walls from inside the room, even though from the outside it looked as if there wasn't anything there at all.

Private Hawk moved over and quickly pulled on

one recessed handle, stepping back and lowering his gun as he did.

The doors opened outward to show a curving staircase, moving up into the dark toward the center of the room.

Cort did the same in the other corner. Same exact result. Another staircase, only this one moving down, circling under the other one.

"I'll bet the elevators are across the room," Bogle said.

"We'll take the stairs," Malone said.

"Afraid you were going to say that," Stanton said.

"Let's go, people. Same groups, same spacing," Malone said. "Don't open the outer doors to the floors unless you have to. Cort, take this position and don't let this door close until everyone is inside."

"One request," Hank said. "Stop at twenty floors to rest and to explore that level. I want to see what one of the higher levels was used for."

"So would I," Bogle said.

Malone nodded. "Twenty floors. Go."

Two of her men disappeared up the stairs, moving silently but quickly.

Malone waited, letting Private Hawk take the lead as they started the long climb to the fiftieth floor and the sky bridge.

13

Time: 2:22 P.M. Pacific Time
12 hours, 51 minutes after Arrival

H ank was panting and sweating by the time they reached the twenty-eighth floor of the alien tower. Private Hawk had his back against the door, holding it open into the large space beyond, and light flooded the stairwell. After twenty flights in the dark staircase, with only their flashlights for illumination, the sudden brightness made him squint.

"What a relief," Stephanie said, as they went through the door and out into the open area.

Hank felt exactly the same way. It was a relief just to stop climbing stairs.

It was cold, with a slight wind swirling through the enormous room. Hank glanced around and could see that the wall and part of the floor below had collapsed in one corner. The outer walls were still transparent like the seventh and eighth floors, but there were more remnants of what might have been furnishings, varied equipment, and patterns of spaces laid out across the vast area.

He and Stephanie moved out into the room and stopped. Hank took a few deep breaths of the wonderfully cooling air, working to get his lungs full again. He had spent far too many years working in labs and not exercising. Climbing twenty stories of stairs spaced just a little too far apart was a killer. And they still had a distance to go, let alone get back down in the other building. He had a feeling that by the time this was all over, his knees would ache for a week.

On the way up he and Stephanie had talked a little about what the size of the stair spacing, the height of the ceilings, and other such details might indicate about the look and shape of this city's original inhabitants. They decided they didn't yet have enough information to make any logical assumptions.

Slowly the other scientists came out of the stairwell, seeming to appear from a door in the middle of empty space. Hank didn't know if he'd ever get used to seeing that. Or to being in a building with invisible walls.

Sergeant Malone stood to one side, watching as they all came up. She and her men were in such good shape that it looked like they hadn't even broken a sweat under their armor after ascending the twenty flights of stairs. And they were all carrying what looked like heavy equipment.

Stephanie reached down and tried to pick up a piece of debris, but it broke apart in her hand. "How old is this city?" she asked, brushing the dust off her hands on her pants.

"Old," Edaro said between pants for breath. "But not as old as I feel at the moment. I haven't had this much exercise since the last time I tried to walk thirty-six holes in one day."

"It's a wonder this entire city isn't just one big pile of dust," Hank said, staring at the broken, crumbled bits of the item Stephanie had tried to pick up. He tried to lift what might have been a piece of furniture, but it too disintegrated in his hands, sending a small cloud of dust swirling into the air. He managed to save a small piece for his pack.

"It's not a pile of rubble," Edaro said, "only because of the incredible engineering of whoever built it."

"We've got movement, Sergeant!" Waters called out from a position near the door.

Hank felt his stomach twist. Aliens from the ship or Sand, he thought. And of course, base hadn't been able to warn them because their communications were cut off when they were inside the building.

"Where?" Sergeant Malone asked, voice calm and all business.

"On the street level," Waters said, pointing at the hole in the wall. "Two blocks off and moving this way from the north. Outside the debris barriers."

Sergeant Malone moved over toward the hole in the wall.

Hank glanced at the other scientists. "Stay here." Then he followed Malone, staying behind her and out of the way.

Malone strode toward the hole in the wall and floor with no attempt to stay hidden. When she got within a few steps of the opening, she stopped. Then she tried to contact base.

"Any luck?" Hank asked after a moment.

She shook her head as he moved up beside her. "Links with the orbital stations are still blocked by the buildings. And we're not high enough for our link to the base to clear the surrounding buildings."

Hank didn't much like the idea that they were out of contact with the rest of the Union and its forces, but at the moment they couldn't do much about it. He looked down through the hole in the wall. The cold air was blowing in at a pretty good pace, whipping at his skin and jacket. In the distance between the buildings, he could see slivers of the blue waters of the Pacific. For some reason the sight of it was reassuring after the last hours in the midst of the alien city.

As he watched, two Union fighters streaked past over the water. That also made him feel better. He might be feeling that they were alone here, but they were far from it.

Below them, the streets looked like narrow canyons and were much farther down than he'd expected from twenty-eight stories. He had a sense that they were climbing more than normal-length flights of stairs. But with the wider-spaced stairs, it was hard to tell. From where they stood it looked more like forty normal human stories in the air. And the sky bridge that was their goal wasn't that far above them. Maybe six or seven alien stories was all.

"Pharons," Malone said matter-of-factly.

"What?" It took Hank a moment to see them. Then he did. They were moving slowly, spread out, coming up the street from the north. Their armor glittered brightly, even in the shadows of the street.

"A dozen warriors, at least," Malone said. "The one in the center looks like a high priest. See that big headdress?"

"High priest?" Hank echoed.

Malone nodded. "That's what our people have named them. I guess because of the fancy headdress and that they seem in command of the troops in battle. From what I saw in the briefing tape, they're the really nasty ones. The others are just soldiers."

"Do they know we're here?" Hank asked.

"Let's hope not. Harden, Marva, with me," Malone said into her commlink. "The rest of you I want up the stairs. Six floors to the sky bridge. Stay together and wait there until I say otherwise. Hawk, at two stories up, break off and shoot some recon views from there, then continue on up."

"Copy that," Hawk said, and fairly dived through the open door.

Hank stepped back out of the way, but he had no intention of going up the stairs just yet. He was going to stay with Malone and see what happened next.

"All right if I stay here?" he asked. "And out of the way."

Malone glanced at him, then nodded.

Private Marva appeared from out of the stairwell carrying the rocket launcher he'd been hauling around since they started. He followed Harden over to Malone at a jog.

Hank watched as Stephanie and the other four scientists, along with five Union soldiers, ran for the stairs as fast as they could go. The door to the stairwell closed and then vanished behind them. Hank spotted a hunk of debris directly in front of the door and decided to use it as a landmark to find the door again quickly.

Then, as he turned around, Malone said, "Damn."

Below, on the street, the band of aliens had stopped, and one of them was gesturing up toward their location. The motion sent a shiver down Hank's spine. He didn't much like the idea of an alien knowing where he was. Let alone pointing at him.

He wanted to step back, to run and hide.

Instead he just stood and watched.

"They know we're here now," Malone said. "Let's sting them before they can do anything with the knowledge."

Hank liked that idea a lot.

"Draco launcher up. I also want the Bulldogs to fire grenades at the same time. Range finders off, aim about fifteen meters over their heads."

Private Marva rested the long, fat tube of the Draco expertly against his shoulder as he loaded a missile in. Hank knew the weapon fired Arrowhead missiles with enough power to blow a hole in the side of a concrete bunker. He'd seen one do just that once during a test. No alien would stand up to one of them.

"Marva, take direct aim on the center of their group. I want that priest out of commission if we can do it. Get them leaderless, and we'll be better off."

Marva nodded.

Hank stared at the aliens. They were fairly spread out across the center of the street. From this height, taking out the one with the bigger headdress was going to be some fantastic shot. He wasn't even sure if he could tell which one *did* have the biggest headdress from so far away.

"Harden," Malone said, "you and I will fire two grenades each, quick succession."

Below them the Pharons were in an open area, giving Malone and her men a clear line of sight as they moved to the opening in the side of the building. And the aliens had a clear view of them in return.

"Okay, people, one attempt at this is all we get," Malone said. "We fire and drop back into the stairwell. If one of their energy beams makes it through this opening, we're all dead. Understand?"

Then Malone turned to Hank. "Get that door open for us and be ready to run up those stairs."

Hank nodded. "Done."

He did a quick run through the debris over to where he knew the door was and faced it as it opened. Then he blocked it open with his back.

Malone nodded to him, then turned back to the hole in the wall, pulling her rifle up into grenade-launch position.

Through the transparent walls Hank could see down into the street and the Pharons below. They had stopped and were gesturing up at them.

"In position, Hawk?" Malone asked into her commlink.

"Fire!" she ordered a moment later.

The Arrowhead missile shot from the Draco launcher with a thump, trailing smoke that swirled up into the room. The grenade launchers made a louder *whomp* sound as Malone and Harden fired an instant later.

The missile exploded among the Pharons in a massive white cloud of dust, hiding them. An instant later the explosion echoed up through the buildings, shaking dust loose as it filled the huge room like an angry wave.

Malone and Harden fired the second round of grenades, then as a unit all three turned and ran toward Hank at breakneck speed.

He waited until they were nearly on top of him, then turned and opened the door to the stairwell, not looking back as he ran, focusing his flashlight on the stairs ahead.

Behind him he could hear Malone and her men following closely, the pounding of their boots like drums in the dark stairwell.

They had all rounded the corner of the stairs when the building shook around them, and Hank could feel his ears popping from the concussions behind them.

The aliens had fired back. And done so fairly accurately.

The sound was deafening as everything seemed to rumble in the surrounding darkness.

Again the air filled with dust as the entire building shook.

Hank didn't even slow down, forcing himself to put one foot after another on the widely spaced stairs, keeping his entire attention on focusing the light just the right distance ahead of him so he wouldn't trip. Falling was the last thing he wanted at that point.

Another explosion—a smaller one—rocked the building just before they reached the next landing. He went through the door, then across and through the next door to the next up staircase, not even slowing down.

His heart felt like it would explode out of his chest, and his lungs were starting to burn. He feared he might choke at any minute on the dust in the air.

About halfway to the next floor level Sergeant Malone shouted to him, "Dr. Downer, let Harden take point. Harden, slow the pace about half."

Hank almost said, "Thank you," then decided he didn't have enough air to speak.

Below them another explosion rocked the building as Harden patted Hank on the shoulder and moved past.

They clearly had pissed off the Pharons.

And Hank was sure that wasn't a good idea.

14

tephanie stopped and tried to catch her breath from running up the stairs just inside the thirty-fourth floor of the alien tower. The sounds of the explosions and the building shaking as they were climbing the stairs had scared her, but she'd forced herself to take deep, full breaths, exhaling completely to calm her nerves. They might have to run again quickly, and she wanted to be ready.

After this expedition, assuming she survived it, she was going to have nightmares for months.

"Are they alive down there?" Bogle managed to ask Private Waters, who had set up his equipment the moment they reached this floor.

"The sergeant and the others should be coming through that door just about now," Waters said.

At that moment that was exactly what happened. Stephanie was very glad when Hank did so, and she rushed over to see if he was all right while he tried to catch his breath. He and the troopers had obviously raced up the stairs hard and fast.

"Where are the Pharons?" Malone asked Waters, making her way over to them, not even seeming winded by the run up the stairs.

"Pulled back two blocks down the street. They stopped there."

"Can you tell how many there are?"

"I can't, Sergeant," Waters said. "Too far for this equipment to show that."

Malone nodded. "They'll be coming back. We just surprised them this time." She turned to Private Hawk. "Got the recon on the fight?"

"I did," he said, turning a very powerful video camera around so the sergeant could see the display. She watched in silence for a moment, then nodded. "Good."

"May we?" Hank asked from beside Stephanie.

Malone nodded.

Stephanie had no desire to see how much damage they had done to the Pharons, but she went with Hank and the others anyway as they went to look at the play-back display on the camera.

"I'll run it in slow motion," Hawk said, clicking the rewind and starting the images.

On the small screen Stephanie could see the Pharon warriors shambling up the street slowly in their ornate, golden armor. It felt almost like being face-to-face with them.

The walking dead.

The Rotten, as Stanton had called them.

Way, way too close. She wanted to turn away and not watch, but she forced herself to stay put. She might need this information.

Their armor was beautiful, covered with symbols and hieroglyphics, and with high, ornate collars behind their heads. They all carried large antique-looking weapons and support tanks on their backs. She assumed those were the tanks Stanton had mentioned that kept the fluids flowing through their bodies.

But their faces were nothing more than wrappings and gray skin, long since dead, as if frostbite had taken it. If what Stanton had told them about the Pharons was accurate, these warriors were in pretty good condition for dead beings. Only one or two had complete chunks of them missing or rotted away.

Just looking at them made her queasy. It wasn't as if she hadn't seen plenty of gruesome sights in her life as a physician, but there was something about the aliens that turned her stomach.

"Who's the one all dressed up?" Bogle asked.

"We call them high priests," Stanton said. "They run the show."

Stephanie focused her attention on him. The high priest's ornate decoration was even more impressive, and he carried two long, curved blades, one in each wrapped hand. He had on a long, mailed skirt that looked like it was made out of some shiny metal. There was an ornate, jewel-like object over his head that Stephanie guessed had some strange purpose or another.

Suddenly on the display the missile streaked in and smashed straight into the middle of the group, about ten meters beyond the priest, exploding in a surrealist sort of slow motion. For an instant the bright flash clouded the image, then it cleared again.

"Yes!" Bogle exclaimed behind Stephanie.

"Good shot!" Hank said.

On the screen it was clear that at least four, maybe more, of the aliens had been blown completely apart by the missile explosion. But the priest had survived.

The next moment, even as the warriors were still turning toward the explosion, two more blasts ripped into the Pharons. One cut the legs off a warrior, while a second explosion ripped off near the priest, sending him blinking backward at an impressive speed.

"We think the priests wear some sort of phase generators," Stanton said. "Allows them to move at very quick speeds if they need to."

"Phase generators?" Edaro asked. "Similar to the technology that's moving this island around?"

"You got me," Stanton said. "Might be."

Stephanie continued watching the small playback monitor as if she were watching a slow-motion car wreck in progress. She wanted to turn away, but she couldn't.

Suddenly two more explosions tore into the Pharons, completely destroying another warrior.

At that moment three of the other warriors took aim upward with their weapons. A seemingly long moment later, they fired while the rest of their kind, along with the priest, pulled back to better cover.

Those shots must have been the explosions that had rocked the building while she'd been ascending the stairs.

"That's it," Hawk said, stopping the playback.

"Took out a bunch of them," Bogle said.

"They aren't going to retreat from this," Stanton said firmly.

Stephanie turned to him, as did Sergeant Malone.

"My understanding, Dr. Stanton," Malone said, "is that they will not break off any attack."

Stanton nodded. "That's right. There will be no stopping them. They will not retreat."

"So how do we fight them," Hank asked, "besides the way we just did?"

"We stay ahead of them," Malone said. "And we pick how we strike and when."

"Or we get the hell off this island," Stanton said, "and let the Union military blow this city right out of the water."

"Why were you even sent along on this mission, Stanton?" Bogle asked, the disgust clear in his voice.

Stephanie had been wondering the same thing.

"That option, Dr. Stanton," Malone said, her voice low and controlled, "has been in place since the moment the Pharons landed on this island." She stared directly at Stanton, then went on. "If we're killed, this island will be destroyed."

Stephanie felt her heart race at the sergeant's words. Stanton's face went completely white, but at least he had the sense to say nothing more.

There was an uneasy moment of silence, broken by Private Waters.

"Sergeant, the Pharons are moving again."

In two steps Malone was beside Waters, staring at the motion sensor. Stephanie desperately wanted to know what Malone was seeing, but she didn't want to get in her way.

She tried to make herself breathe slowly, get her pulse back to normal, just as she'd learned to do back in the emergency-room days of her intern years. The ER staff needed nerves of steel so they could think and act quickly.

Finally, Malone looked away from the motion sensor and clicked her commlink. "Listen up, people," she said. "They're coming into this building below us."

"Damn," Bogle said.

Malone went on. "We've got a thirty-four-story head start, and we're going to use it to get to that energy source before they do."

"You're assuming they were headed there," Edaro said.

Stanton laughed. "It doesn't matter where they were headed," he said. "From what we saw in the Cache, they're persistent and vicious. They won't give up until we're dead."

"Exactly," Malone said. "I'm counting on just that." She turned to Waters. "Let me know the minute they're on the second floor."

Waters nodded. "Just entering the building now."

She pointed to the open corner of the building that led out onto the sky bridge. "When I give the order we move across that bridge quickly, keeping spread out. We get to the next building, then ascend what looks to be about twelve floors to the next sky bridge leading to that building over there."

Stephanie's gaze followed where Malone was pointing. Then Malone turned back to them. "Harden, Vasquez, bring up the rear. I want enough explosives on that sky bridge to take it down on my order. Understood?"

Harden and Vasquez both nodded.

"You're going to try to blow the bridge with them on it?" Hank asked.

"Exactly," Malone said. "But what I'm going to damn well make sure of is that they can't follow us over those barricades down to that energy source."

Right at that moment Stephanie was very, very glad that Phoebe Malone was in charge of the mission. Malone was one of the most competent people she had ever met. And that was saying something.

"The Pharons are on the second floor," Waters said.

"Let's move it," Malone said. "Jenkins, take point. The rest follow me. Single file, move quickly once you're on the bridge. We don't know how solid it is."

Another thought Stephanie didn't want to contemplate.

She moved quickly behind Hank over to the edge of the large room where the corner of the building led out onto the wide walkway to the next building over.

"Oh, shit!" Lee said, his usual smile completely gone.

"I don't think I can do this," Edaro said, twisting

the golf ball in his hand like he was trying to wring water out of it.

Stephanie agreed. She didn't think she could do it either.

The sky bridges, when looked at from the side, through the clear walls of the building, looked contained, just as the walls of the buildings had looked solid from the street below.

But from the inside of the bridge, it became very, very obvious that it, too, was built of the same material as the building's walls.

It was completely transparent.

Ceiling.

Walls.

And *floors*.

From the inside, looking across at the hole in the building on the other side, there didn't look to be anything at all in the air between them.

From the inside, the sky bridge was completely transparent.

"How crazy were these aliens?" Bogle asked.

Stephanie watched as Jenkins took the lead, seeming to walk right out into thin air, moving as surely and as quickly as if he was walking across a solid field of dirt. Two of his comrades followed him, acting as if there was nothing in the slightest odd about all this.

Stephanie could feel her head spin. If she'd had anything to eat lately, she was sure she'd have thrown it up right there.

"Okay, people," Sergeant Malone said, "just don't look down."

"No kidding," Hank said.

"So which way do we look?" Stephanie asked, not trying to be funny.

"I can't do this," Lee said. "Sorry."

"I'll be fine," Hank said.

"I'm not sure I can do it, either," Stephanie said.

"Just close your eyes," Hank told her.

"Good idea," Malone said. "Marva, help Dr. Lee across. Cort, you help Dr. Peters. Doctors, close your eyes and don't open them until I say so."

Lee nodded and did as she ordered.

Stephanie also nodded.

"Got you all the way, Doc," Private Marva said to Lee, heading out into what looked to Stephanie like sheer emptiness.

"Ready?" Cort asked, taking her arm firmly.

She took a very large and deep breath, then closed her eyes and focused on the feel of the hard floor underfoot, imagining a concrete highway. "I am."

Cort's grip felt reassuring as they started off. And not once did the ground feel anything but solid under her feet.

And not once was Stephanie tempted to open her eyes.

15

**Time: 2:49 P.M. Pacific Time
13 hours, 18 minutes after Arrival**

C rossing that invisible bridge thirty-four alien
stories in the air behind Stephanie and Private
Cort was the hardest thing Hank could ever remember doing. Every step of the way he was
convinced he would plunge to his death.

Yet with each step his heel hit a hard surface, a
surface he couldn't see. After about ten steps of looking down at the street far below, he decided he'd better keep his eyes focused straight ahead, at the
building they were approaching.

Also, with every step he was convinced the Pharons
would come swarming out of the building below and
literally blow him out of the air.

Yet they didn't appear.

He felt exposed.

Deathly afraid.

And dizzy.

One step after another, as fast as he could make himself go, he followed Stephanie to the other side, finally gaining the seemingly more solid ground of the large room in the new building.

"That was no fun," he said.

"We there?" Stephanie asked.

Private Cort laughed. "We're there, Doc. Okay to open your eyes." He made sure Stephanie was facing inward, away from the ramp.

She let out a huge breath, as if she'd been holding it all the way across, and slowly opened her eyes. She glanced around, then smiled. "Thanks."

Cort nodded and moved off.

Stephanie turned to Hank. "You all right?"

He could feel the sweat dripping down his face, but he was okay, now that he was back on a surface he could see under his feet. Amazing how visual humans were.

They turned around and watched as Kelly Bogle, his tall frame moving slowly, came across the invisible bridge, followed by Stanton and Sergeant Malone. It was odd, watching them walk on what seemed to be thin air. It was like a bad movie effect. Only this was real.

Too real as far as Hank was concerned.

"So what were the builders of this place thinking?" Stephanie asked. "They must have really loved the wide-open air, that's for sure."

"Seems that way," Hank said, glancing around at the transparent walls of the floor they were on. "And they had no fear of heights."

"Lucky them," Stephanie said.

Bogle was also sweating when he finally reached the safety of the new building.

"Find the stairway and get the door held open," Malone ordered a pair of her men.

Hank watched as two more of Malone's people came across, then said, "There's no place to hide explosives on that walkway. How do you plan to handle that?"

"Shaped charges aimed down the sky bridge," she said. "Placed right here." She pointed at the opening to the bridge. "Enough to blow a hole in the far side of that building over there."

Hank only nodded. With luck, that might take a lot of the Pharons out. If nothing else, it should slow them down some.

"Okay, people," Malone said, "start up the stairway to the next sky-bridge level. Keep a good pace. Twelve stories."

Hank turned, then he and Stephanie made their way through the ruins to the stairwell door now visible and open.

Behind them Malone watched as her two men started setting the explosives to blow up the bridge.

Fifteen very long minutes later, they emerged onto the forty-sixth floor of the building. It was similar to the one they had just left, with debris scattered around and another sky bridge leading off the west corner.

Waters had used an open area near one side of the room to set up the motion detector.

Hank went right over to him. "Where are the Pharons?"

"Moving slow," Waters said. "Lead warrior has about thirteen more floors to go. They're scattered over three floors."

Hank nodded and moved over to where Stephanie stood, staring out at the sky bridge below, which the Pharons were going to have to cross to get to this building. There was no other bridge, so if they didn't come across there, they would have to return to the ground level and start up through another building.

And that would slow the aliens even more. Maybe long enough for Hank and the others to get to the energy source inside the debris barrier, whatever it was.

Edaro and Dr. Bogle joined them, followed a few moments later by Stanton and Lee. Hank was startled at how differently each of them was taking the experience. Edaro seemed almost unfazed, his golf ball an almost constant companion. Bogle seemed focused and serious, his face always set in a contemplative frown. Lee always looked like he was enjoying himself, while Stanton sweated so much, Hank was amazed he had a drop of liquid left in him. Hank was afraid, but nowhere near as much as Stanton was.

Bogle looked down at the sky bridge. "Looks so damn solid from up here."

"Even painting a line on the floor might have helped," Edaro said, setting up his equipment.

"Maybe there were lines there when this city was inhabited by the people who built it," Bogle suggested.

"More than likely," Hank said. "I'm not looking forward to crossing the next one, that's for sure."

And he wasn't. The next time he just might close his eyes, put his hand on one wall, and walk until he tripped on something in the next big room. Far less traumatic on his nerves.

"I thought it interesting," Edaro said. "Not something we're used to by any means, but imagine working in a building like this and going out for a walk next door."

"Don't even talk about it, would you," Lee said. "I'm getting dizzy again just thinking about it."

Edaro laughed and flicked on his instruments for detecting the molecular stability of the island.

"It's been over two hours since the earthquake," Stephanie said, crouching beside Edaro. "We due for another?"

"Of course we are, eventually," he said, again. "But let's just hope we're off the island when it happens, since I'm betting the next one will shift this place to a new location."

"I don't think I'll take that bet," Bogle said.

Hank didn't like the sound of that at all. At the moment they were a long way from their departure point.

"And we wouldn't survive that, would we?" he asked, thinking he already knew the answer.

"I doubt it," Edaro said. "If what I think happens actually does happen, it would more than likely short-circuit every electrical impulse we have in our bodies."

"So how do the Sand do it?" Bogle asked.

Edaro shrugged and glanced at Stephanie. "You have an answer for that?"

She chuckled, then said, "Give me one for about six months of tests and I might. No, make that a year."

"Hey, look at that," Edaro said, surprise filling his voice.

Hank knelt beside Stephanie and looked where Edaro was pointing. It was a shallow spike on the running graph.

"That spike," he said, "occurred right at the moment we attacked the Pharons."

"Pharon priest phasing back out of danger?" Hank asked.

"They use a phasing generator to help them move around quickly," Stanton said. "So that could be it."

"More than likely," Edaro said. "Which, if true, means that the power that shifts this city around the Maelstrom is very, very similar to the phasing power the Pharon priests use. Only magnified a hundred thousand times."

"So now maybe we know why the Pharons are here," Bogle said.

"With them," Stanton said, "it's always unhealthy to impute motives other than death and fighting."

Sergeant Malone came over to stand looking over the shoulder of Private Waters. Hank turned so he could hear what she was saying.

"Make sure we blow it before any of them get across," Malone said.

Waters nodded. "Three warriors in the big room now."

Malone spoke into her commlink. "Get set, people. On my command blow the bridge."

Hank glanced back at the other civilians. "About time for fireworks," he said.

They all gathered closer to the edge where they could see the solid-looking sky bridge below them. Hank had no idea how something that looked so solid could be so invisible from the inside. A level of light control that was far above anything humans had reached, that was for sure.

"Four on the sky-bridge level," Waters said. "First one at the sky bridge."

"Wait until they start across," Malone said.

"First one starting across now," Waters said. "Another right behind the first."

"Blow it now!" Malone ordered.

Hank saw the explosion before he felt it. Dust shot out into the air from the sides of the sky bridge as if under high pressure.

Then the building they were in began to shake. Dust fell from the roof like a fine mist as the concussion from the explosion traveled up the building.

At first Hank didn't think Malone's people had used enough explosives to destroy the bridge. But then, below him, as if in slow motion, the bridge separated from their building and swung back toward the one holding the Pharons.

Then that side also broke loose as a massive cloud of dust filled the air, pouring out of the building like smoke from a five-alarm fire.

"Oh, man," Bogle said. "That's got to hurt."

Hank watched as the sky bridge broke apart on its

long fall to the street below. When it finally hit the street, the sound must have been deafening. Forty-six stories in the air, Hank couldn't hear it at all at first. Then a low rumble shook them, like a distant clap of thunder.

The dust continued to pour from the opening in the building where the Pharons were. He watched it for a moment, then turned and went to join Waters and Malone.

"Success?" he asked.

"Took out the four on the sky-bridge level," Waters said. "And from what I can tell, another just inside the staircase. The rest are starting back down."

"So how many does that leave?" Bogle asked.

"Ten of them," Waters said.

"Too many," Hank said.

Frighteningly, Malone nodded in agreement and said nothing.

16

ergeant Malone watched as the readings on the motion sensor being held by Waters showed the Pharons moving back down the stair core of the building across the street. Blowing the sky bridge had taken out five of them and bought her a little time. But nothing else. And just then she and her people were sitting ducks for that Pharon ship if the priest decided to call it in for some kind of air attack. She had no idea if the Pharon ship had weapons or not, nor what kind of cover the Union forces could give against the alien ship.

Either way, she didn't want to wait around to find out.

Up there, forty-six stories in the air, they were too much out in the open, easy for a Pharon ship to target.

And she had to get everyone across and down fast.

"You're sure none of them made it across into this building?" she asked Waters.

"Positive, Sergeant," he said.

"Good. Keep an eye on that motion detector. I want to know if anything moves anywhere around us. Any chance of it spotting the Pharon ship?"

"No," Waters said, "not unless it comes within three blocks and hovers. More than likely it will come in too fast for me to give you any good warning."

She nodded, turned, and faced the civilians. "Okay, we've got another sky bridge to get across. And we have to do it fast."

"Problem, Sergeant?" Dr. Downer asked.

So far he'd been the most levelheaded of them all, for which she'd been thankful. So she didn't mind answering his questions, keeping him working along with her instead of against her.

"We're in a vulnerable position up this high," she said. "I want us across and headed down toward that energy source as soon as we can get there."

He nodded.

"Down into the area where the Sand creatures are," Dr. Stanton said.

"That's right, Doctor," she said, turning to glare at him directly. She hated weakness, and he was obviously weak and afraid. "Should be a good chance for you to study an alien race up close and personal."

Stanton paled visibly, and she turned back toward Dr. Downer, who looked as if he just might laugh.

"Two by two across the bridge. Jenkins, Vasquez, take point. Make sure the down stairway over there is clear."

Without even so much as a "Yes, Sergeant," the two turned and headed toward the corner of the room that connected with the sky bridge.

"Follow them, people," she said. "Move it, move it, move it."

She waited until all six of the civilians had started across. Dr. Downer with Dr. Peters, Bogle and Edaro together, Private Marva again with Dr. Lee, and Dr. Stanton making it on his own just fine.

Walking on a transparent bridge hadn't bothered her past the first few steps on the last one, so she used the time crossing this one to study the sides of the building toward which they were headed, as well as the barricade below them. It was a perfect vantage point. But far too vulnerable to air attack.

She studied the barricade. The building they were headed toward was inside the barricaded area. And there were two other intact-looking sky bridges above her, leading from it to other buildings also inside the barricaded area. But none led to nearby buildings outside of the area. Good. That meant if she could plant some explosives on this bridge and if the Pharons were stupid enough to try to follow them again, she'd make them pay again.

She did a quick study of the sky bridges she could see from there. She was going to have to get these people out from the barricaded area at some point. And doing it this way had worked out fine so far. If they blew up this bridge, there were others to the south that would lead them out.

She studied the condition of the skyscraper toward which she was walking. It had two holes in the side, one about two stories below, forty-four alien floors off the ground. She examined it as best she could, noting that it was large enough to shoot through, but small enough to make a hard target from the street below.

There was another hole in the building just five stories off the ground. Too close, too open. The one two stories down just might give her a chance at one more attack from above on the Pharons. It looked out over the street and onto the front of the building the Pharons were in. Another Arrowhead missile just might cut them down to size a little more.

"Waters," she said into the commlink, "location of Pharons?"

"Between the tenth and fifteenth floors, moving down."

There was time. A good Arrowhead missile and maybe a few well-tossed frag grenades just might do some more damage. And at this point, if they ran up against those Pharons face-to-face, she wanted the number of aliens down substantially.

She had no problems attacking where and when she could. She'd be damn stupid not to take any chances she got to cause casualties among the Pharons. Now, just maybe, she might have one more.

She strode across the big room to where the scientists were standing, clearly trying to recover from the sky-bridge ordeal. "No time to stop here," she said. "I want you to take the stairs down two by two. Move quick, but don't try to run. Stop on the tenth floor unless you hear otherwise."

"Let's go," Dr. Downer said to Dr. Peters, turning and leading her toward where Vasquez had the door to the stairwell open. The other four civilians followed.

She turned her back on them and looked back out at the sky bridge while clicking on the commlink. "Vasquez, Waters, and Marva, flank the civilians down the staircase. Hawk, take point. Jenkins, bring up the rear. Stop and secure the tenth floor."

She turned to Cort just as he was coming across the bridge. "When everyone's across, I want shaped charges set here to blow the bridge. Then get down to the tenth floor."

Cort nodded, put his pack down, and started setting the charges.

"Harden, Raynor, you're with me at the opening two floors down. Get the Draco launcher ready."

Malone didn't expect any answers from any of her men, and she got none. What she did expect was action, and that was what she was getting.

Harden and Raynor were coming at a run across the bridge, the last over.

Waters had the motion sensor out so he could track any movement around them, secured against his chest. That left both hands free as he set off with the doctors.

Malone followed the civilians down the stairs to the forty-fourth floor. It was cold, with a good wind swirling through the interior. She grabbed a hunk of debris and shoved it in the door to prop it open, then strode over to the hole she'd spotted. There was a good chance they were going to have to make a quick escape.

Private Harden was right behind her, with Raynor right behind him.

"First Pharon nearing the second floor," Waters said over the commlink. "The others are spread up four floors."

Down in the canyon below her she could see the opening in the building where they had gone in, and where she would wager the Pharons were going to come out. She could also see three of the four corner doors of the building if they came out there. There was only one way out of that building she couldn't see, and it was on the opposite corner.

Harden moved in beside her, the launcher cradled in his arm. "Ready."

"Good. Think you can put one right through that hole in the bottom of that building?"

Harden snorted. "Give me something hard, Sergeant."

She turned to Waters. "I want to know when two of them are on the ground floor."

"One reaching it now," Waters said over the commlink. "Another not far behind."

"On my mark," she said to Bradley.

She watched that hole like a cat watching a mouse hole. An Arrowhead missile exploding in the big, open, ground-floor area of that building ought to take out anything in the room, as well as do damage up the ramp system to those above. At least it was worth another shot. They'd only brought four Arrowhead missiles, but she figured this was worth the loss of one.

"Two on the ground inside," Waters said. "Another on the ramp coming down."

"Do it," Malone said.

The next instant the Arrowhead was streaking at the hole. Harden's shot was perfect, the missile almost seeming to duck under the top edge of the hole in the wall.

The next instant the hole looked like a dust-and-smoke geyser. The explosion's muffled sound and the faint rumble from it reached them a moment later.

"That's got to smart," Raynor said.

Suddenly Malone caught a glint of something reflecting light out of the corner of her eye. About fifteen blocks over and above them.

The Pharon ship. Looking as fearsome as it was alien, it had a swept-down nose and was gold and black. It was like no aircraft she'd ever seen or imagined.

Two Union jets were on attack runs at it from above. But the alien ship was coming at them.

"Move it, move it, move it!" she shouted. "Into the staircase. Fast!"

She turned her back on the hole and took off at a run.

Harden and Raynor were a half step behind her.

If she wasn't mistaken, that had to be the craft that had brought their Pharon friends to the island. Or maybe even a second one, for all she knew. And she had no idea how it was armed. But simply standing there and waiting for more information would surely cost them all their lives.

"Freeze and cover in the stairwells," she ordered

into the commlink as she ran. "Get the civilians down and covered."

She was still twenty steps from the blocked open door. It felt like a kilometer.

"Cort, where are you?"

"Six stories below you," his answer came back as she reached the door and waited for her men to get through before kicking the debris out of the way.

As the door swung closed she saw the Pharon ship headed straight at them. The two Union jets were never going to reach it in time, even if they could stop it.

Then she was through the door and down into the dark stairwell, following her men. The doors had just closed behind them when the explosion rocked the building.

It smashed into her back like a full kick, sending her tumbling with the debris, kicking her light loose and plunging the stairwell into blackness around her.

She managed to roll about ten steps, her helmet and suit taking the brunt of most of the impacts before she came up on her feet and slammed against a wall with her right shoulder.

Maybe she could brace there and ride it out. She pressed herself against the shaking and rumbling wall, working to get a purchase with her feet.

She managed to hold in that position for almost a full two seconds in the dark as the stairs and building around her shook and roared.

Then something massive smashed into her waist, sending her tumbling downward again with what seemed like an avalanche of debris.

She was bounced and jarred like a kid's doll in the mouth of an angry dog.

Head over heels, her rifle flew from her grip.

The world around her seemed to explode.

She could feel one leg break as she smashed into fallen rubble, but she felt no pain.

Then something heavy smashed against her chest.

Huge.

Thick.

Too much to shove away.

It pinned her to something rough and hard, ripping at her Kevlon suit as if it were tissue paper.

The last thing she felt was the pressure as whatever was on her chest rolled.

Up her body.

Over her head, smashing her helmet like a thin-shelled peanut.

And sending her into the blackness that every good soldier knew might someday take her.

17

Time: 3:27 P.M. Pacific Time
13 hours, 56 minutes after Arrival

The explosions from above felt to Hank as if some giant had taken the entire building and just shaken it. Hard. And for a very long time.

Hank and Stephanie had first gotten the warning that something was about to happen from Private Vasquez.

"Get down! Cover!" Vasquez shouted from the stairwell below them.

They were on the landing of the twentieth floor, so when he heard the warning, Hank pulled Stephanie over against the staircase wall, and both of them covered their heads, huddling together. Hank's light was between them, Stephanie's pointing upward over their heads.

The next moment a massive explosion shook the building, knocking them both to their butts. Hank dropped his rifle and held Stephanie close to him with one arm while keeping his head protected with the other.

The roaring and rumbling went on and on as the building shook.

Pebbles and small chunks of the walls pelted down on them like a hailstorm. Dust choked the air around them, getting into Hank's eyes. He just hoped nothing bigger than pebbles and dust let loose over them. Or that no large part of the structure below them had collapsed. This building was so old, anything was possible.

The floor under them jerked upward at one point, bouncing them both hard enough to hurt.

Then slowly, the rumbling stopped.

Hank could barely see Stephanie's face through the swirling dust, his beam of light nearly useless.

He coughed, then said, "You all right?"

"I think so," she said. Her voice was shaky.

He felt the same way.

What had happened? It didn't have the same feel as the earthquake, and Vasquez wouldn't have known it was coming to warn them. More than likely, this was some sort of counterattack by the Pharons.

He pulled them both to their feet, then picked up his rifle and slung it over his shoulder. Their backs were against the wall, but other than the choking white dust, that was all he could see.

He knew that Lee and Bogle had been about a half a floor above them, and Stanton and Edaro another half a floor above that when the explosion hit. Soldiers had been between them, and Vasquez had been just ahead of him and Stephanie, a dozen steps down the staircase. Clearly the explosion had been up a lot of floors higher. He just hoped everyone had gotten out alive.

Around them the dust was settling quickly. His light beam now could reach the other wall. "We've got to see if everyone is all right," he said.

Vasquez came up out of the dust like a ghost, completely white and coated with the fine powder. Hank glanced at Stephanie and then at his own arm. They both looked the same way.

"We've got wounded up higher," Vasquez said. "Lost track of the sergeant. Looks like the Pharons hit back."

"Damn," Stephanie said.

Hank said nothing. He didn't want to think about Malone being killed. More than likely her commlink had just been damaged. And until he knew otherwise, that was what he was going to believe.

"Dr. Peters and I can work our way back up," Hank said. "We'll see if we can help with the wounded."

"Good," Vasquez said. "I'll take the others and secure the tenth floor as the sergeant ordered."

"You got an extra commlink?" Hank asked before Vasquez could move.

"Waters does, two flights up."

"Ask him if he could have one ready for me when

we get there. And see if he can let one of your men stick with us."

"Got it, Doc," Vasquez said.

"One more thing," Hank said. "Who's next in the chain of command of this unit?"

"Private Cort."

"Thanks."

"Let's go," Stephanie said, turning and heading back up the stairs into the dust.

"Move slow and steady," he said, securing his rifle better over his shoulder so he could have both hands free. "You don't want to be gulping too much of this dust into your lungs."

"Good point," she said.

Hank followed her as they climbed.

At the next landing they ran into Bogle and Lee, with one of the soldiers, starting down.

"You three all right?" Stephanie asked, shining her light at each of them.

"Shook up," Bogle said. "A mass of bruises, but otherwise fine."

"Good," Hank said. "We'll meet you on the tenth floor."

Bogle nodded, and the three of them moved off into the dust-choked air while he and Stephanie continued upward.

Stanton and Edaro, with Privates Waters and Hawk, were on the landing to the twenty-second floor. Stanton was sitting, his back against the wall, as Edaro used a first-aid kit to cover a cut in Stanton's forehead. Stephanie knelt and did a quick check of the cut, then nodded.

"You'll be fine."

"How about the headache?" Stanton moaned.

Stephanie just patted Stanton's arm and stood.

Hank glanced at Edaro. "Any chance that was another shift earthquake?"

"No chance at all," Edaro said.

"Afraid you would say that," Hank said.

Private Waters handed Hank a commlink and helped him get it on so that the earpiece fit and the microphone was against his cheek.

"Can you get them down to the tenth floor?" Hank asked Private Hawk.

Hawk nodded and reached to help Stanton to his feet.

"Who's above us yet?" Hank asked Waters.

"Cort and Jenkins have reported in. They're heading back up to see if they can help Sergeant Malone, Harden, and Marva."

"No word from them at all?" Stephanie asked.

"No, Doctor," Waters said.

Stephanie shook her head sadly. "Not good."

Hank agreed with that completely. They were in an alien building and had just been attacked by aliens. Now they'd lost the leader of their mission. Not good was an understatement.

Hank quickly keyed in the commlink. "Private Cort, this is Dr. Downer."

"Go ahead, Doctor." Cort's voice came back strong.

"Dr. Peters and I are headed back up to see if anyone needs medical attention. That all right with you?"

"Fine, Doctor. Thanks. I want everyone else continuing with the sergeant's orders until we get back in touch with her. Waters, are you with Dr. Downer?"

"I am," Waters said.

Hank heard Waters both over the commlink and in person. It was a strange stereo effect.

"Stay with Dr. Peters and Dr. Downer."

"Copy that," Waters said.

"What are you doing?" Stanton asked, looking first at Hank, then Stephanie.

"We're going back up," Stephanie said, glaring at Stanton through the dust. "There may be wounded up there, and they might need my medical help."

"Keep them safe getting to ten," Waters told Hawk, then turned and waited for Hank and Stephanie before starting up the stairs.

The dust slowly cleared as they climbed up the next fifteen flights of stairs, moving steadily. Saying nothing.

Hank could feel the dust clogging his nose and caking the inside of his throat. He just hoped that whatever this alien building was made out of wasn't toxic to humans, because they were swallowing a lot of it.

At the landing of the thirty-seventh floor, Hank stopped them and took a drink from his canteen, then clicked the commlink on.

"Private Cort?" Hank said. "This is Dr. Downer. What's your position."

"Private Cort here, Doctor. We're on the landing of the forty-second floor."

"Any sign of the sergeant or the others?"

"None," Cort said. "The staircase is blocked. We're trying to dig our way through."

"Understood," Hank said. "We'll be there in a few minutes."

Stephanie took a long drink from her canteen, then put it away. "Five more floors."

"Let's do it," he said.

It ended up taking them a lot longer to go the last five floors than it did the first fifteen. Massive hunks of wall material had fallen into the staircase in a number of places, forcing them to crawl and climb over it. By the time they reached Cort's position, Hank felt more like he'd climbed a mountain than flights of stairs.

The moment he saw what Cort and Jenkins were digging at, just above the landing on the forty-second floor, he knew they hadn't a prayer of getting through. The stairwell had been completely jammed full of debris.

Cort and Jenkins had made some progress clearing an area out of the top, but not enough. It would take a crew of fifty days to go two floors through that stuff, and even then he doubted it would make a dent.

Stephanie did a quick check of both the slightly built Jenkins and the dark-haired, wide-shouldered Cort and pronounced them fine.

"Private Cort," Hank said, "do you mind if I try to find another way up?"

"Not at all," Cort said. "Waters, stay with them. Report if you have any luck."

Hank was impressed. Even with the sergeant only missing, the chain of command seemed to have immediately clicked into gear. Cort had simply stepped in as if he'd been in charge the entire time.

Hank led the way as they entered the large room. After being in the dark, dust-filled stairway, with only flashlights for light, going into the bright, open space was a shock. There was dust in the air there, too, but not as much.

A large chunk of the ceiling had collapsed on one side, and Hank could feel a draft, meaning there was now a hole in the side of the building somewhere on the other side of the collapsed ceiling.

He moved directly across to the other side of the big room. As he approached where he knew the building's second utility core was, a door opened, showing him a dark, dust-filled room.

"Would you cover us from the door, Private?" Hank asked. "And keep it blocked open."

"Not a problem," Waters said.

With Stephanie behind him, Hank moved inside, clicking his flashlight back on as he went. When the group had first figured out that the alien buildings had two utility cores, Malone had decided they should stay in the one with the stairs, never exploring the other. But he was betting this one had some sort of lift or maybe even a second staircase that might not be blocked.

This level looked exactly like the one they'd explored on the other side. He moved to the corner where the up staircase, if there were one, should be, and pulled on the door. It opened to show exactly what he was hoping it would. A staircase. The cores were identical. Mirror images of each other.

He shined his light up it through the dust. There was a massive amount of rock and debris on the stairs, but it looked passable.

He clicked the commlink. "Cort, I think I may have found another way up. Directly across in the other utility core."

"Copy that," Cort said.

"Stay here," Hank said to Stephanie, who was standing with him on the landing. Cautiously he climbed up ten steps, keeping his light above him to make sure nothing was about to come loose and fall on him.

"Don't go too far, Doctor," Waters said. "Cort is on the way."

"Think you might be able to climb up there this way?" Stephanie asked from below as he eased his way up another few steps.

"Possible," he said. "Better chance than the other stairwell at this point."

At that moment Cort and Jenkins joined them.

"Let's go easy," Private Cort said. "If I make it to the next floor, I'll call for the rest of you to follow."

"Understood," Hank said.

He watched from his position as Cort disappeared around the corner of the staircase, climbing over piles of debris, knocking some rock-sized hunks loose.

A long minute later Cort said through the commlink, "I'm on forty-three. Come on up."

Hank glanced at Jenkins.

"Single file," Jenkins said.

Jenkins turned and led off. Hank followed him ten steps back. The debris was crumbling and loose, but he managed to scramble over it and make it to the forty-third floor.

Together, he, Cort, and Jenkins pushed open the jammed door into the main room and blocked it open with a hunk of debris. Meanwhile the others also arrived up there.

Hank was shocked at what he saw through the door. The forty-third level was completely destroyed.

A blast had cleared the floor of all debris, smashing it against the windowlike walls on the far side. A gigantic hole in the ceiling on one side of the building showed clear air up two more stories. Whatever had hit on the floor above had blasted through the walls and three floors of the building with more force than Hank wanted to imagine. In fact so much of the structure was gone, he was amazed the upper part of the building hadn't just tumbled over.

"Were they up there?" Stephanie asked, pointing at the remains of the floor above them.

"They were," Cort said. "Question is, did they have enough time to take shelter."

Hank moved across the cleared floor to where the other staircase was. The door didn't open as one of them usually did, but between him, Cort, and Jenkins, they managed to get it braced open enough to get inside.

It was there, right at the foot of the staircase, that they found Sergeant Malone and Private Raynor.

Both had been smashed under the hunks of ceiling and walls and rolled down the stairs. The sergeant's helmet and upper torso stuck out into the open, but Hank knew it wasn't necessary to see if she was alive.

Her helmet was an unnatural shape for any human head.

Private Raynor was pinned from the waist down under a massive hunk of wall just beside the sergeant.

Stephanie went to him first and quickly checked for a pulse. Then she glanced over at Malone and stood, shaking her head.

"They're both dead," she said.

18

Time: 4:12 P.M. Pacific Time
14 hours, 41 minutes after Arrival

Stephanie reached the tenth floor, moved out into the light, and sat down on the floor, her back against the invisible wall around the utility-core stairwell. She couldn't remember the last time she'd been so dirty and tired.

And so discouraged.

It felt as if she'd spent her entire life on this island. But it had been only a little more than six hours.

Six hours and they'd found no tech or other useful-looking artifacts for the Union, but they certainly had information about two alien races.

Six hours that had cost three good people their lives.

They'd found Private Harden's body just above Sergeant Malone's and Private Raynor's, six meters farther up the stairs. It was obvious that they hadn't stood a chance in the massive explosion from the Pharon attack.

And there was nothing the group could do at that point but leave the bodies. She hated that more than anything else. Private Cort put a subsonic marker near Sergeant Malone's body, just in case someone could come back and get them. That way at least they could find them again.

But that didn't make it any easier to turn away and start down the stairs, leaving them there.

And now the mission situation was critical. Six civilians, six Union soldiers, confined in a building where the only hope of escape was to descend into the area where the Sand creatures were. Trying to reach the sky bridge was no longer an option, since both stairwells were blocked. They had to go down into the part of the city that was barricaded off from all the rest, including their escape route to the water.

Granted, the entire point of going up into the buildings had been her idea to get them into the barricaded area. But now, with Sergeant Malone dead, trying to reach the energy source no longer seemed like such a great idea. If it was up to Stephanie, they would get away from the city and off the island. Immediately.

But there was no turning back.

They had to go down, no matter what might be waiting for them.

Only four of the remaining six soldiers were in the

large room on the tenth floor. Cort had assigned the others to new positions on the floors below.

Sitting across the floor in a clear area were Bogle, Lee, Edaro, and Stanton. They all looked lost, just as she was feeling. Stanton was almost shaking. And there was no sign of Edaro's golf ball or Lee's smile. Only Hank seemed to be picking up the slack and trying to keep going. He was crouched alongside Cort and Waters, watching as they worked on the motion detector. It, too, had taken damage in the explosions. Their satellite uplink to the orbital stations had been destroyed, so the only way to contact the mainland would be to get closer to the water or to rise high enough to get a commlink signal out of the tall buildings.

In essence, they were on their own for the moment.

And the ocean and their escape were on the other side of the debris barricade.

The light coming through the invisible walls was very dull, and the debris in the room cast deep shadows. They had very few hours of daylight left. And if Edaro was right, she doubted they had many hours until the island shifted again, taking all of them along to who knew where.

She worked to catch her breath for a minute, watching Hank. He was covered in dust, hadn't eaten, and had to be as tired as she was, yet he hadn't slowed down. It was a good thing Major Lancaster had put him in charge of the civilians. He would keep them going and would work to save all their lives as well as the mission.

And Cort had stepped in with total professionalism. She studied what she could see of him behind his standard armor. He had black hair, intense black eyes, and wide shoulders. He spoke slowly and carefully and treated his Special Ops comrades with respect. With Sergeant Malone gone, the group was lucky to have him as their leader.

Still, Cort and Hank were going to need as much help as they could get if any of them were going to make it out alive. Maybe it was time to see what she could do.

She pushed herself to her feet, took another deep breath, and walked over to the four doctors.

"Dr. Edaro," she said, "did your molecular-sensing equipment survive the explosion?"

"I don't know," he said, pulling his pack around and digging into it.

She watched as he quickly set up the odd-looking device with the metal bars poking out of the back. After a moment he glanced up at her. "It's working."

"Good," she said. "Any idea how long we have?"

He studied the screen for a moment, typing in a few commands, then shrugged. "From what I can tell, the energy buildup is continuing. Right now it's far greater than it was before the earthquake four hours ago."

"Greater?" Lee asked. "Not good."

Edaro nodded slowly while staring at the screen on his device. "Nothing about this is good."

"You didn't answer my question, Doctor," Stephanie said, even as she feared his answer. But they needed to know. Hank would need to know. "How long?"

Again Edaro just shrugged. "What exact level of energy will cause the island to phase to a new location is anyone's guess. That will totally depend on what the trigger is. All I can do is guess that it will happen sometime in the next three or four hours, assuming a constant rate of buildup of energy."

"Great, just great," Stanton said.

"So it could happen any minute?" Hank asked.

"It could," Edaro said. "But again, it depends on the trigger, on what exactly causes the phasing. I could give you a better guess if I knew the answer to that."

Hank nodded. "Okay, maybe we'll get you that answer, but first we've got to get out of this building."

"And how do you propose we get past the barricades?" Stanton asked.

"Same way we got in," Hank said, glancing at Stephanie. "Just not the same route. But we've got to try and make one quick stop first."

"The energy source?" Bogle asked.

"If we're going to bring back anything from this mission, it's got to be there," Hank said. "Let's get moving." He turned and headed toward the staircase.

Stephanie followed, quickly catching up.

"You got the motion sensor working?" she asked as they headed for the stairwell door.

"Working like a charm," Hank said. "Private Waters has it. Nothing in the building below us at this point."

"The Pharons?"

"In a building on the other side of the barrier."

That was the first piece of good news Stephanie had heard in the last hour. At least they would still have a warning if someone or something was coming at them. That would help some.

Hank stopped at the staircase door and waited for the other four civilians to catch up with them.

"Where exactly are we going?" Edaro asked.

"We're going to find that energy source," Hank said, "by the most direct route we can find."

"We're dead," Stanton said softly.

Stephanie desperately wanted to just backhand the pale-skinned excuse of a man, but it wouldn't have done the mission any good, no matter how good it would make her feel.

"Not just yet, Doctor," Hank said, his voice cold.

"I don't feel dead, do you?" Bogle asked, reaching over and pinching Stanton hard on the arm.

Stanton twisted away, but said nothing more.

Hank glanced around at Bogle and Lee. "Better check your rifles to make sure they're still working."

Stephanie touched the pistol in her pocket as the others checked their guns. So far she hadn't needed to take her pistol out of her pocket. But it was still there if she needed it.

Hank turned to Private Cort. "We're ready when you are."

Cort nodded. "Do it like the sergeant would have, people," he said into his commlink. "Two-by-two cover from the second floor on. Vasquez with Jenkins, Marva with Waters. Move on my command. Hawk, behind us."

Stephanie was impressed with Cort. And with this

entire unit. They had just lost three of their comrades, including their sergeant, and they were still functioning like a clock.

She just hoped the soldiers were impressive enough to keep them alive. She had a feeling that their only hope of getting out of the city while there was still time was the six Union soldiers.

19

Hank could see the debris barricade just one block behind them and about five blocks ahead of them in the other direction. The effort was crude, but the barricades were obviously intended to protect the center of the city from penetration. Not a very effective defense, which only made him wonder all the more why it had been done. What was in this area of the city that someone, at some point in the past, thought important enough to try to protect? Was it the power source? Or might it have been something else? There was no way of telling at the moment.

The air was cold around him, and the light from

the Maw didn't even come close to illuminating the deep canyons of the streets. Every detail was cast in long, black shadows, making it seem closer to night than four-thirty on a clear afternoon.

From what the motion sensor told them, nothing was moving on the streets around the building. And there was no longer any sign at all of where the Pharons had gone. At last sight they had been climbing up the levels of a tower a few blocks away. Hank could only hope they hadn't made it in here yet.

Private Cort had tried to contact base via the commlink when they left the building, but nothing seemed to get through. And he got no response at all.

The energy source was another block deeper into the barricaded center of the city.

And down.

It was the down part that had him the most worried. He didn't at all like the idea of going underground. Not in the slightest. But if they were going to salvage the mission, they might have to. And do it fast.

"Move out," Private Cort said into his commlink.

Hank watched as the four soldiers leapfrogged forward up the street on both sides, two giving cover while the other two ran ahead.

"Stay against the wall," Cort ordered Hank and the rest. "And keep a good three meters distance at all times."

Cort nodded for Hank to follow Private Jenkins along the wall behind the front four.

Hank had his rifle up, loaded and ready to fire. He carried it in one hand, leaving the other free as he moved along the side of the building behind Jenkins. Stephanie followed behind.

It didn't take them long to go the block.

Nothing tried to stop them. If they were to meet resistance, it would most likely be as they started into wherever the energy source was.

Or maybe they would get lucky and meet no resistance at all. Hank knew that was dreaming.

The building that seemed to contain the energy source was the shortest by far in the entire central city. Surrounded by hundred-story skyscrapers, it looked almost tiny at fewer than twenty.

Like a four-pointed cap, sky bridges led off in four directions from the top of the building. Clearly this had been one very special building to the original residents of this city. And Hank had a sneaking hunch they were just about to find out why.

They moved to the corner across from the edifice, and Cort had his men take up cover positions. From there Hank could see three of the sky bridges leading from the top of the energy building, but not the other structures around it. Maybe, if they had to, they could just go up inside the energy building to one of those four capping sky bridges.

They'd be taking a chance, since he didn't know if they could get up and across the barricade that way or not. At the moment, they would have to play their escape by ear until they had more information. He didn't think that was much of plan, but it was the only one they had.

"Everyone ready?" Cort asked.

Hank glanced back at the other scientists and gave

Stephanie a smile, then turned back to Cort, and said, "As ever."

"Get that door open," Cort ordered his men.

Two of them were already there near the corner of the building. On Cort's order, one reached up and pushed on the corner. As with all the other buildings so far, the corner split and opened inward.

Privates Hawk and Marva ducked inside, one going right, one left.

"Clear," Hank heard Hawk say. "It's a small room with corridors leading both left and right."

"Copy that," Cort said. "Marva, go right. Hawk left. Check it out. Vasquez, Waters, cover them."

Then Cort glanced around at the scientists. "Move it, people. Through the door double time."

With Jenkins in the lead, Hank ran across the street and into the open doors of the building.

Just as he entered, he heard Hawk say, "Clear up here, but you've got to see this."

"Same this way," Marva said. "Flat amazing."

The room was small, like an entryway. As Hawk had described, two wide corridors veered to the left and right, along the outside wall, which was not transparent. Vasquez covered one hallway, Waters the other.

Cort went left, and Hank followed him down the corridor the way Hawk had gone. For a moment he thought they were going to need flashlights, but then he saw light twenty steps ahead. And the closer they got to where Hawk was pressed against the outside wall, staring into that light, the louder became a humming sound in the background. It was as if the area ahead was almost alive.

Cort moved up beside Hawk, also pressed against the wall, rifle ready.

Hank did the same.

And what he saw through that door made him forget where he was for the moment.

The entire twenty-story building was hollow and maybe four times as big as the insides of any of the other buildings they had been in. There were ramps circling up around the inside of the building, and from the fifth or sixth floor to the sky bridges, the entire building was basically transparent. He could see all the other skyscrapers towering into the blue sky above. This place was far, far larger than any indoor, covered football stadium on the mainland.

Bigger than anything he'd ever been inside.

"Spectacular," Stephanie said beside him.

Granted, the hollow building was something to behold, but it was what was below them that Hank couldn't get his gaze away from.

The building was hollow down a good ten more floors into the ground. At the moment they were standing on what was basically a wide ledge. And square in the center of the massive open space below was the biggest machine he had ever seen.

It had to stand three stories tall and would have covered a normal New York City block, yet from inside it almost looked small. It was lined with some sort of bright, shiny material polished brighter than any chrome Hank had ever seen.

Around the massive machine were thousands of Sand, moving in a zombielike state. All of them wore the black robes that hid their faces, hands, and feet.

There seemed to be no purpose at all to their movement except to stay close, without touching, to the massive machine.

He had no idea what they were doing, or why. He was just glad that at the moment none of the Sand seemed to notice him standing above them.

The hum that reverberated throughout the space was coming from that machine, which made it seem alive.

"One of you doctors want to tell me what that is?" Cort asked, turning to Hank.

Hank glanced back at where the other five also stood staring into the massive, hollow building and the giant machine it held. They seemed almost in shock at the sight.

"Dr. Edaro," Hank said, "is that the energy source that's powering the island phasing?"

Edaro seemed to shake himself, then yanked his backpack around so he could get to his equipment. He quickly had it out and working. Within seconds he glanced back up at Hank. "That's it."

"If I were a betting man," Bogle said, "I'd bet that was the biggest phase generator in existence."

"Could be," Edaro said, nodding.

"It's similar in exterior design to what the Pharon priests use to speed them around a battlefield," Stanton said. "Only about a million times bigger."

"The energy field around it is building from underneath," Edaro said. "More than likely some sort of massive storage down in the ground there."

"Can we get this recorded?" Hank asked, focusing them back on their task at hand. "I want every measurement you can take on this thing as quickly as you can take it—photos, energy readings, molecular cross sections. Everything."

"Going to have to make it quick," Cort said. "We got Sand coming at us up the ramps from below."

Hank glanced over at where Cort had indicated. Six or seven Sand were seemingly floating up the ramp, moving slowly and in one group toward them.

On the other wall six more were headed up.

"Can you hold them off?" Hank asked.

"For a time," Cort said.

"Buy us some," Hank said. "Taking all the information from this that we can get might just make this entire mission worthwhile."

Hank watched as Cort deployed his men, two each along the tops of both ramps leading down to the floor holding the machine. He left Private Waters guarding the door from the street.

Beside him Lee, Edaro, Bogle, and Stanton went to work, using the equipment they had brought to record every detail, both seen and unseen, of the alien machine. Hank desperately wanted to take them right down to it, but with the thousands of Sand below, milling around on the lower level, plus the ones coming up, there was no chance of that happening.

The information they could get would be as good as they could get. He just hoped it was enough to justify the lives of three good Union soldiers.

"Amazing," Edaro said. "Stanton is right. I just did a single molecular cross section of that machine. It does look like some sort of massive phase generator. Extremely advanced."

"But why the hell would it be in the middle of a city?" Stephanie asked.

Hank glanced up at the buildings towering over this one. "I'd bet this machine was working for a long time before any disaster hit this city. In fact, I'd wager the whole place was mostly built around this building. It just has an older feel. And a central feel."

"It does, doesn't it?" Stephanie said, nodding as she looked around.

"So why build a massive phase generator here, then build a city around it?" Stanton asked as he worked a video-recording unit.

"What better way to explore than take your entire city and culture with you," Edaro said.

"You're saying these people were a race that jumped around through space in their cities?" Bogle asked.

"It's kind of logical," Stephanie said.

"Could be," Lee said. "Fascinating structure for a culture, that's for sure."

"I still don't think it happened that way," Bogle said.

Hank looked up at the sky bridges above his head, then pointed. "This building was here before many of those other skyscrapers were built. The center of this city built up around this place, not the other way around. You can tell by how the architecture fits."

"Learn that in physics class, Doctor?" Stephanie asked, smiling at him.

"Logic class," he said.

"I tend to agree with Hank," Edaro said, not looking up as he worked. "You don't start up something this big in the middle of a city as an experiment. I'm sure this was a tried and trusted form of technology before the city was even built."

"So what happened?" Bogle asked.

Edaro shrugged. "Any of a thousand things. A plague and no one turned off the lights. Maybe the city got caught by the Maelstrom on one jump and sucked in and that caused the destruction."

"So what happened to all the inhabitants?" Lee asked.

Bogle pointed. "Maybe we're looking at the few survivors right down there."

"Speaking of Sand," Cort said, "I've still got a dozen of them coming at us. How much longer?"

"Need ten more minutes," Edaro said, "to get a complete set of molecular cross sections of that machine. I'm taking them every few meters. We get that and we can build one if we want to."

His words excited Hank. Maybe the mission might just be a success after all, if they could escape with such revolutionary knowledge. He turned to Private Cort. "Guess we're going to have to stop them."

"Better find some cover from the winds," Cort said.

Hank didn't like the sound of that. But at that moment, as the Sand slowly came up the ramps toward them on both sides and the air around them got colder and colder, it looked like very good advice.

20

Time: 5:02 P.M. Pacific Time
15 hours, 31 minutes after Arrival

Stephanie was ordered to a spot in the corridor leading to the outer door just before Private Cort gave the order to fire on the Sand advancing up the ramps. From there she could see the approaching creatures, plus part of the floor below covered with them.

It would be a slaughter of the Sand on the ramp, but she knew it had to be done. The closer they came up the ramp, the colder it got, as if the Sand actually sucked the heat energy from the air. She couldn't imagine what would happen if they touched a human, but she didn't want to have to find out.

Just seeing them down there had shocked her.

She'd heard the descriptions of them from the first battle, but actually seeing the creatures milling about in their black robes startled her. It was as if seeing them was the one piece of evidence her mind needed to accept the fact that they turned to sand when shot.

Now it looked as if she was about to see that event for herself, firsthand.

Around them the air was getting colder and colder. Her breath was freezing in front of her, and she could feel the cold pulling the heat of her body through her jacket. How cold was it on that floor around the machine ten stories below? She couldn't imagine.

Hawk and Marva opened fire on the group coming up from the left.

Waters and Vasquez on the ones ascending the right ramp.

The shots echoed off the walls of the thirty-story-tall room. The Sand went down into piles of black robes, dotting the wide ramp like the mounds of a burrowing animal.

Below, around the big machine, the other Sand seemed oblivious to the sound of weapons fire. None even tilted their heads to look up. They just kept milling around the monster phase generator that filled most of the center of the gigantic building.

Stephanie watched as the piles of robes on the left ramp seemed to dissolve into piles of black sand. It had happened. Before her very eyes they had actually become sand. All of her medical training protested against what she had just seen. Yet her eyes had seen it.

And not once in any of those piles did she catch

even a glimpse of a body part like a stray arm or leg. Yet the soldier's bullets had downed them.

Then, after only an instant, the robes were completely gone, also became just sand.

And then the cones of wind started from each one.

Small at first, then they got bigger and bigger, becoming small tornadoes, sucking the black grains into the air.

Energy was being released from each body. Seemingly massive amounts of it. And she had no idea what, how, or why it was happening. In all her professional career, she had never felt so totally uneducated. She didn't like the feeling.

"Back to shelter! Now!" Cort shouted.

The four soldiers who'd fired scrambled back into the hallway with the rest of them. At her feet Dr. Edaro broke off doing his molecular cross-sectioning of the large machine to hold on and ride out the expected coming winds.

The funnels of wind over each Sand grew and grew, picking up the remains of the bodies in black swirls, climbing higher and higher in the huge space inside the building.

One story, then two stories tall.

Then five stories up the winds swirled.

The growing monsters of whirling winds yanked at her jacket, pulling at her as if the Sand wanted her to join them. She held on to the wall, watching as best she could as dust from the ground and walls around her pulled upward.

The piles of Sand were now completely gone.

The winds pounded her one way, then the other,

with almost bruising force. The temperature around her had dropped by at least fifty degrees and the winds seemed also to want all her body heat, just as the Sand had.

Beside her Hank fought to keep the commlink secure on his head with one hand while holding on to his rifle with the other. He had his feet braced wide and his shoulder against the wall.

On the ground below her, Edaro bent over and put his entire body over his machine to protect it.

Stanton fought to remain in place as he tried to video what was happening.

Slowly the funnels of winds started to merge, growing bigger and bigger still.

Below them, the wandering Sand seemed totally unaffected. Their robes didn't move, even though dust swirled around them.

Finally, all the funnels had joined into one big, swirling trail of black sand, its tip rising twenty stories over her head.

Then, as if being pulled, the top of the funnel seemed to turn toward the large machine, pouring downward like water from a fountain.

As she watched, the downward motion of the leading edge of the black stream sucked the rest of the funnel up and then back down behind it, like a long tail following a body.

The leading edge of the wind smashed head-on into the massive machine, releasing the black sand and dust onto the floor in front of it.

There the black sand re-formed into a pile of robes.

And the pile of robes stood up and moved away, joining the others.

A dozen of them re-formed, the same number that had been shot.

Then there was no more wind in the monster room.

Just silence and the hum of the machine, much louder now.

The dozen resurrected Sand were no longer at all distinguishable from the rest of their kind circling aimlessly on the floor below.

Stephanie and the rest of the scientists and soldiers all stood and just stared for a long instant. Stephanie wasn't even sure she was breathing.

The air around them was quickly warming up. She tried to swallow and just couldn't. Her throat was too dry, her mind too confused.

Then Lee said, "Someone want to explain to me what the hell just happened?"

Stephanie looked at Hank, who only shrugged. From all the laws of physics and medical science that she knew, what she had just seen wasn't possible.

Period.

But neither was the Earth being in the Maelstrom, so maybe some of those rules should now just be tossed away. Actually a lot of them.

"I hope someone recorded that," Hank said softly. "Because no one back at the facility is going to believe us. Hell, I don't believe it, and I just watched it."

"Got it all," Stanton said, holding up a miniature camera.

"So did I," Lee said, holding up another miniature

camera that fit inside his palm. Stephanie hadn't even noticed he was filming.

"Good," Hank said.

"Tell you what, people," Edaro said, staring at his screen. "Let's try not to shoot any more of them."

"Why?" Stephanie said, kneeling next to where he sat on the ground with his machine.

Edaro pointed at a spike on a graph that filled the screen of his weird-looking computer. "Because by doing so," he said, "we just fed the energy supply below."

He looked up, first at Stephanie, then at Hank. "It means we have less time until the entire city phases and takes us with it."

"How much less?" Stephanie asked.

"Significantly less would be my guess," Edaro said. "Those creatures pumped a lot of energy into that thing."

"Damn," Hank said softly.

Stephanie looked down at the milling Sand and their monster machine. She knew it was crucial that they take this information back with them. They had come too far, paid too high a price to not get it back now. Who knew what Union scientists could do with what they were getting here? Future phase generators on armor maybe.

Or quicker transport to any place on the planet.

Or stronger building materials even.

But if the city jumped with them, none of that would happen.

She looked up at Hank. "We've got to get out of here."

Hank nodded. "How soon until you're finished, Chop?"

"Three minutes," Edaro said.

Then Hank turned to the others. "Have we gotten everything we can possibly get from here?"

Everyone nodded, so Hank turned to Private Cort. "Up or back outside to get out of here? What are your plans?"

"If we go up," Cort said, pointing at the ramps, "we will be exposed on those building faces."

Stephanie shuddered at the thought. From what she could see, the ramps around the sides of the building above about five floors were all transparent. The only reason they were even visible was that the bottoms were like normal wall material. And the sides of the building along each ramp were transparent.

She studied the room above her. Each ramp circled exactly halfway around the building to climb one floor. To climb the fifteen floors, they'd have to walk around the inside of this building seven and a half times on transparent ramps, then cross a transparent sky bridge.

She could do it, but she wouldn't like it.

"Yeah," Hank said, "but outside we take the chance of running into the Pharons. And the only way to cross that barricade is to cross over it from above."

Cort nodded, looking up and studying the ramps.

"I don't much like either choice," Bogle said.

"I agree on that one," Stanton said.

"I think we'd be better off making a run for it up the ramps in here," Hank said. "And try to reach those sky bridges as quickly as we can."

"Why?" Cort asked.

I get it," Stephanie said. "The Pharons, more than likely, have been trying to get something from this place for years. We have to assume they understand what killing a Sand does to the energy."

"More than likely," Hank said. "So you think they might not want to fight in here if they show up?"

"After what just happened," Stephanie said, "do you?"

"No," Cort said. "We go up." He clicked on his commlink. "Vasquez, Hawk, get up that left ramp. Spread out and cover us as you go. We're taking that way out."

"This is going to be rough," Lee said, his smile gone.

"We'll get you up there, Doc," Private Cort said.

Lee only nodded.

Stephanie looked up at the transparent building walls and the transparent ramps and shuddered again. What had she been thinking?

At least they were heading back.

That thought alone could get her over a lot of transparent ramps.

She hoped.

21

T he ramps that sloped at a gentle angle up the inside of the massive, hollow building were over six meters wide, with no railings of any sort. They were as wide as a large living room, yet for Hank's taste, they weren't wide enough by far. Especially at the fifth floor, where they became transparent.

And the exterior walls also became transparent.

It was like stepping out into empty space.

"A ramp to heaven," Bogle had said.

Hank found it disturbing to suddenly be moving up a transparent ramp along the side of a transparent wall high over the heads of the mob of milling black-robed Sand.

Within ten steps out onto the transparent part of the ramp, he felt his hands sweating, and he could barely move. Yet they all had to keep going. There were still a good fifteen floors of this above them.

Privates Vasquez and Hawk were almost to the top to set up cover positions.

Hank kept moving another hundred paces, then stopped and looked back.

None of the troopers seemed to be bothered by this strange ramp in the slightest. Or more than likely, their training was so good, they just didn't show fear. Just then Hank wished he had the same training.

He half leaned against the solid, but transparent outer wall, feeling as if he were actually standing in open space. It was amazing to him that after all the years of wear and tear, these ramps and walls could remain so perfectly transparent. Or more than likely, what had been surface markings on the transparent surfaces had worn off. Either way, it made for one very disconcerting experience.

Stephanie was behind him, and he waited for her to catch up shakily, letting the solid feel of the wall and the ramp under his feet reassure him. Her face was white, and she was having problems breathing. There was no chance she was going to make it all the way to the top. None at all.

Not this way at least.

"We're going blind," he said to her. "Neither one of us will make it any other way."

She looked at him like he was mad.

"Put your hand solidly against the outer wall."

She did as he told her.

"Now close your eyes and tell me what you feel under your hand and feet."

"Hard wall, hard ramp," she said.

"Exactly. Keep your eyes closed until I tell you to do otherwise. Don't take your hand off the wall, and you'll be just fine."

She nodded without opening her eyes.

"I'll be right in front of you every step of the way. So trust me; I'll tell you if you need to open your eyes again. Okay?"

"Thanks," she said. "This is better."

"Don't start yet," he said. He looked back down to where the other scientists were having their own problems negotiating the transparent ramp and wall. Lee seemed unable to move at all. He had frozen after about ten steps out onto the clear surface, and Cort had ordered Waters to help him.

"Close your eyes, Lee!" Hank shouted back at them. "Walk with your hand along the outside wall as a guide. It helps take away the visual problems."

Lee nodded and did as Hank suggested.

So did Bogle.

So did Edaro.

"I'll bring up the rear," Stanton shouted. "This doesn't bother me."

"Great!" Hank shouted back. "Dr. Lee, just walk smoothly and quickly along the wall, tracking it with your hand. You'll be fine."

Lee only nodded without opening his eyes.

"Let's move out, people," Cort said, motioning at

Hank. "Marva, stay behind Dr. Stanton and make sure nothing comes up at us."

"Understood," Marva said over the commlink.

Hank didn't wait to watch how they were doing. Instead he glanced at Stephanie, seemingly standing in midair beside him, her eyes closed, her hand resting against the transparent wall. It was a very weird sight. If it hadn't been so damned scary, it would have been funny.

He let the feeling of the solid wall under his hand give him confidence. "Ready to walk?"

She nodded. "Ready."

"Then let's go. Take it at whatever pace you feel comfortable with. We have some climbing to do before we reach heaven."

Stephanie laughed as she stepped toward him hesitantly. "Solid, brown dirt would be heaven right about now."

"That's exactly where we're heading," he said, staying just ahead of her.

Slowly she picked up momentum. Behind them the others were doing the same.

They were all moving again.

Hank kept his hand firmly on the transparent wall as they moved upward, his eyes closed for ten of every twelve steps. It was enough to keep him focused on the solidity of the ramp.

Enough to keep him going.

Twenty of the longest minutes of his life later, he stood beside Stephanie at the top of the ramp, looking back down at the massive phase generator and the myriad Sand milling around it. From that height the

Sand almost looked like tiny ants. And none of them had tried to follow, which was good.

He was sweating.

She was pasty white and sweating. But at least they had made it this far.

The rest of their group looked as if they were walking on air coming up at them.

Around them Hank could see the towering sky-scrapers and the four sky bridges leading across to them. The sky bridge to the south looked fine, and ten stories up higher in the building to which it led was another sky bridge leading even farther south. With luck, it would be far enough to get over the barricade.

But from here he couldn't tell.

"South sky bridge looks good to me from here," Hank said into the commlink.

"Be right there," Cort said.

A few seconds later Cort cleared the top of the ramp behind Dr. Bogle.

"Jenkins, Vasquez," Cort said, "take up point across the south sky bridge and secure the stairs over there."

"We've got to go up another ten, I'd guess," Hank said into the commlink, looking up.

Cort nodded.

Bogle smiled ruefully at Hank, the sweat dripping off his face. "That was a real fun-house ride."

"The builders of this place were real sadists," Stephanie said between gasps of breath.

"They just had no fear of heights, loved beautiful views, and enjoyed climbing stairs and ramps," Hank said. "I can think of much worse."

"They would have been interesting to meet," Stephanie said. "But it doesn't seem—"

"Shit," Cort said, glancing at the motion sensor he now carried around his chest. "We've got company."

"Pharons?" Hank asked.

"Looks like it. They're coming in over the west sky bridge. Looks like eight of them."

Hank didn't know what to say. The west bridge just wasn't that far from where they stood at that moment. The entire sky-bridge floor was like a giant ring above the generator below. And the four bridges were like four spokes off that ring, equally spaced apart. They had come up onto the ring between the south and west bridges.

"Run for it!" Cort ordered. "South sky bridge. Dr. Bogle, get Dr. Lee across that bridge. Quick."

The entrance to the south bridge was a good hundred paces away, and it was going to take at least another hundred running paces to get across it. And two of their men were still a short distance down the ramp and racing upward as hard as possible.

They were all going to be sitting ducks the moment those Pharons got across the bridge and onto the ring.

Without another look at the opening of the west sky bridge, Hank turned Stephanie toward the south bridge and they took off running.

"We need to slow the Pharons down!" Cort said, as he and Waters took up sprawled positions on the transparent surface, rifles aimed at the west sky-bridge entrance, covering everyone else.

Marva was the last up the ramp and dived to

sprawl out beside Cort, also aiming at the entrance as Stanton ran for the bridge behind Hank.

Hank glanced over his shoulder. From the position of Cort and his three men, they had open shots at the mouth of the west sky bridge, but not all the way down it.

And that same angle was keeping the Pharons from firing at them at the moment. But the blockade would quickly go away.

"Dr. Downer," Cort said over the commlink, "would you ask Dr. Edaro if he thinks we dare fire a few grenades?"

"Edaro?" Hank shouted to the scientist running about twenty paces ahead of him across the sky bridge. "Any danger of using grenades back there?"

"Who knows!" Edaro shouted back as he ran. "But don't kill any Sand."

"I don't see how we can *not* fire," Hank said to Cort over the link. "Edaro says just don't kill Sand."

"Understood," Cort said. "Vasquez, get those civilians up those stairs."

Hank reached the other side of the sky bridge and stopped, taking up a position against the right wall, where he could watch what was happening behind him. Everyone else kept running toward the staircase as Cort had ordered. Stanton went past him without even a slight hesitation, open fear in his eyes.

Cort, Marva, and Waters all lay on the transparent surface of the ring, waiting, rifles aimed at the west sky bridge.

Hank had no doubt that it would be only seconds before the Pharons would appear.

"On my mark," Cort ordered, "I want three frag grenades on that sky bridge. Then we get our asses out of here before they have a chance to fire back."

All of them stood.

"Get ready," Cort said.

Hank aimed his rifle down the sky bridge and at the entrance to the west bridge.

"Now!" Cort ordered.

As a unit all three took one step forward and hit the bridge entrance with three perfect lobs.

Without waiting to watch, they turned and sprinted across the transparent surface of the south bridge toward Hank.

Hank stood his ground, rifle aimed beyond them, covering them. Luckily he couldn't see anything to fire at yet.

He hoped the grenades would slow the Pharons down. Otherwise, Cort and his two men would be sitting ducks on the bridge.

The concussion from the grenade explosions knocked the three soldiers to the hard, invisible surface of the bridge. For an instant Hank thought they were falling the full twenty stories to the street below, but then they rolled and came back up running, without even seeming to break stride.

"Marva, Dr. Downer," Cort ordered. "Covering fire."

Hank did as he was ordered, firing one shot after another across the sky bridge, even though he saw no Pharons.

Beside him Private Marva, his dark hair slick with sweat and dirt, was doing the same. The noise was deafening, but it was nothing compared to the blood pounding in Hank's chest and his labored breathing.

He just kept firing. Spacing his shots, until the clip was empty. Then he jammed in another and kept at it.

"Waters," Cort shouted about fifty steps from the staircase opening, "cover Marva and Dr. Downer! Marva, Doctor, swing wide out of the line of fire and get your asses over here."

Hank was on his feet and took three steps out of the way to his right before Waters opened up, firing back down the tunnel just as he and Marva had done. Hank had no idea if Waters could see something, or was just firing as he had done, but he didn't stop to find out.

He just kept running, leaping over and around debris.

Fifty paces. Thirty paces. Twenty more steps and he was at the entrance.

Cort was behind some debris near the blocked open door, rifle up, ready to cover his men.

Hank went through the doorway, got down low, and stuck his head and rifle back out to try to help lay down covering fire.

"Waters, get back here now!" Cort ordered, firing down the tunnel as Marva came running on the right, out of the direct line of the bridge, and took up a position next to Cort.

Waters rolled and came up on the left, also running out of the line of fire.

On the other side of the bridge Hank could see that one Pharon was down, but his golden armor was still moving. Probably just wounded in some fashion or another.

Two other Pharons had weapons up, ready to fire.

Hank aimed at the one on the right and emptied his entire clip at the figure.

Cort fired at the one on the left.

The noise was fearsome.

The smell of the gun firing clogged his nose.

His focus was only on the Pharons.

On killing them.

He emptied a second clip and replaced it, firing again.

The Pharon Hank was aiming at spun around, obviously hit, but then got ready again to fire.

Hit, but not killed.

The damn Pharons were hard to knock down, that was for sure.

Hank emptied another clip at the gruesome-looking thing, then ducked back inside the door to load a new clip. And get out of the way as Marva dived through.

Waters was right behind him, rolling across the floor.

Marva took Hank's previous position and started firing.

The next moment Cort took a running dive at the door as a blast of flame exploded in the large room, sending Marva rolling backward and hurling Hank back against the wall.

The wind was knocked out of him.

An instant later Cort came diving through the

door. His jacket was singed and smoking, and his face looked slightly burnt, but he was alive.

"Up the stairs!" Cort shouted. "Quick. Marva, hit them with a couple of grenades. Keep 'em guessing."

"Got it," Marva said just as another energy-weapon blast lit up the room outside the stairwell.

Cort tried to push himself up, but his legs didn't seem to be working.

Hank grabbed hold of Cort and pulled him to his feet. Waters took hold of Cort on one side and Hank supported the other, then the three started up the stairs as Marva fired grenades at the advancing Pharons with his Bulldog.

They had made it up almost twenty steps in the dark when the first grenade went off, followed by five or six more in quick succession.

"That ought to hold them off for a minute," Marva said over the commlink.

"Let's hope so," Hank said. "We're going to need all the minutes we can get."

"I don't think you know how true that is," Cort said. "Those damn things are hard to kill."

At the moment all Hank could think about was carrying Cort and getting up the stairs.

Fast.

This was the first firefight he'd ever been in. He had a sneaking hunch it wasn't going to be his last that day.

22

Time: 5:43 P.M. Pacific Time
16 hours, 12 minutes after Arrival

The Pharon high priest strode across the sky bridge as his soldiers ahead of him fought the creatures back and into the stairwell of the nearby building. Those creatures, their smooth skin disgusting to him, were proving hard to kill. That annoyed him. He hated anything that stood in his way.

He killed anything that stood in his way.

The creatures had slowed his progress and destroyed beyond saving many of his soldiers, and for that they would pay. This entire planet crawling with them would pay.

"The creatures are moving up the stairwell in the neighboring building."

"How many did you kill?"

"None."

He did not like the answer, but at the moment he could do nothing about it.

He moved through the smoke from the battle to the edge of the ring that looked down over the central area of the large building.

Far below him, his goal stood. The phase generator, large enough to transport this entire asteroid. It had been a long time since he had last stood there. The city had jumped hundreds of times since he had first discovered this treasure. He always escaped only barely before the city jumped, but had come away with enough information to know that recovering the secrets of the massive machine below would advance the cause of his people.

And his place in the rank.

So now he had returned.

Finally.

The sustaining fluids pumped reassuringly through his body as he studied his goal. Thanks to the smooth-skinned creatures' annoying interference, once more he only had a short time to learn the generator's secrets. But there was still enough.

He turned to his soldiers and ordered four of them to follow the creatures. "Kill them. Bring me their bodies to use their fluids."

The four, without a response, turned and headed after the fleeing creatures. Those fair-skinned animals had destroyed too many of his soldiers for his liking.

And their puny aircraft had bothered his ship almost from the moment he had landed. But he had kept his focus on his goal, which now lay before him.

He motioned for three of his warriors to follow him down toward the massive phase generator below. He had very little time and much to learn.

And when he had gotten what he came for, he would take the fluids of those fair-skinned creatures and destroy this planet.

23

"Get everyone across the sky bridge and hold for us there," Cort said into his commlink as they reached the fifth landing.

Vasquez came up and took over for Hank in supporting Cort, who seemed to be slowly regaining sensation in his legs. But not enough yet to walk on his own. Cort was also burnt in a number of places, but the burns weren't life-threatening as far as Hank could tell. Stephanie would be able to do something when they caught up with her.

They paused on the landing just long enough for Hank to get the motion detector off Cort and hand it to

Marva, who had come up behind them. From what Marva reported, four Pharons were coming after them, now starting up the stairs five flights below. Four others had split off and were headed down the ramps toward the lower level and the giant phase generator. It didn't look as if any of them had been killed in the fighting. But Hank knew he had hit one pretty good.

"I just hope they don't go shooting Sand too fast," Cort said, when Marva gave him the update. "We've got to get off this damned island first."

"Couldn't agree more," Hank said. "Come on. Dr. Peters will take a look at you in about five floors."

Vasquez and Waters supported Cort between them, Hank trailing ten steps behind. Marva held back twenty paces, then followed.

By the time they reached the sky-bridge level and crossed to where the others were waiting, the Pharons had made it to the fourth floor below them. Mummy-wrapped and encased in armor, the Pharons were slower than they were, but not much.

And they had a ship out there somewhere. Hank just hoped the Union planes were keeping it busy. Damn busy.

Stephanie took one look at Cort being carried across the bridge and rushed up to meet them the moment they were off the transparent surface.

She did a quick scan of Cort, then gave him a shot. "Painkiller," she said. "Won't knock you out."

"Thanks, Doc," Cort said, but didn't ask what was wrong with his legs. Hank was thinking the guy must be made of metal. He was impressed.

Marva glanced down at the motion detector. "Alien clearing the fourth floor below us."

"We've got to get down lower," Cort said. "I'm betting it was the Pharon ship that took out Sergeant Malone. And we're high enough here for them to do it to us, too."

"I agree," Hank said. "Let me see if I can figure out the best way to go."

He turned and at a fast run headed across the large room toward the far side of the building, dodging in and out of debris as he went.

"Stay with the doctor, Marva," Cort ordered over the commlink. Hank wanted to say there was no point. He wasn't going very far, but he knew Cort was just doing his job. And doing it right, even though he was injured.

There was no point in their going down if they were still inside the barricade. And he bet they weren't out of it yet. The best way to find out was just to look. When he reached the east edge of the building, he could see almost directly down to the street through the transparent wall. The barricade was one more block over to the east, but right below them to the south. Three floors up was a sky bridge to the next building south.

He scanned the sky beyond the tall buildings. No sign of the Pharon ship or any Union planes.

He flipped on the commlink. "Three floors up and across the bridge and we're out of the barricade."

He turned back toward the stairwell at a full run, just as Marva caught up to him.

"Let's move out, people," Cort ordered. "Up the stairwell three floors."

By the time Hank got there, Vasquez was waiting for him and Marva, giving them cover.

It took them only five minutes to scramble up the three flights of stairs and across the sky bridge. But it seemed like it was a very long five minutes.

As Hank ran across the transparent sky bridge, he could see the massive pile of debris blocking the street far below. Finally, they were outside the barricade. He could feel the relief. Being inside the barrier had made him feel trapped.

But at the moment they were still far too high in the air. The lower they got, the safer they would be from attack by the Pharon ship.

Ahead two men were still basically carrying Cort. They all stopped at the staircase door as Cort ordered Waters and Jenkins down to secure the stairs.

Behind them, Marva said that the Pharons had crossed the first sky bridge and were heading up the last three floors. Hank couldn't believe they could almost keep up with them. But they were.

"Those guys are faster than I thought they would be," Bogle said.

"And we still have a lot of floors to get down," Stephanie said. "Down is sometimes harder than up."

Hank knew that.

"Jenkins, I'm starting the civilians down," Cort said. "As fast as they can go. Stay ahead of them."

"Copy," Jenkins said.

"Go, people," Cort ordered.

Quickly, Bogle, Lee, Edaro, and Stephanie passed

through the door into the staircase and then down. Hank hesitated, waiting to see if he could help any.

"We're going to have to make a stand here," Cort said, leaning on Marva. "Buy us some time. Maybe even blow that sky bridge. Hawk, set the shaped charges to take out the sky bridge. Vasquez, give him cover."

Both nodded, and Hawk ran for the bridge entrance, digging explosives out of a pack as he went.

"May I suggest you get down the stairs, too," Hank said. "You're going to be slow."

Cort started to object, then nodded, knowing Hank was right. He glanced over at where Vasquez covered Hawk. "You two about ready?"

Hawk finished setting the charges and ran back to take cover beside Vasquez. "Ready."

Suddenly the staircase door on the other side of the sky bridge opened and two Pharon soldiers moved out into the open. Even from this distance, Hank could see the beautiful golden armor and huge, sparkling jewels embedded in them.

And the gray, rotting faces.

Dead faces.

He had his rifle up and ready to fire.

"Don't shoot," Cort ordered.

Hank held his fire, though he wanted to pull the trigger more than he'd ever wanted anything in his life. Those rotting faces would haunt his nightmares for a long, long time if he ever got out of this alive.

The two Pharon foot soldiers advanced toward them, moving slowly but steadily.

"Frag grenades down the bridge," Cort ordered. "Then open fire."

Cort motioned for Hank to help him through the door into the staircase.

Hank put his rifle over his shoulder and did as Cort asked.

"Now!" Cort shouted, as he and Hank cleared the door.

Hawk and Vasquez stood and threw at the same moment.

The grenades rebounded once off the wall of the bridge, and like good pool shots, bounced at the two Pharons on the other side before exploding.

The concussion hurt Hank's ears, but he ignored it. He watched the dust rising in the other building, filling the space.

"Pump them in there," Hank said.

One after another, Hawk and Vasquez lobbed grenades onto the bridge.

Explosion after explosion followed.

The smoke and dust were pouring out of the near side of the bridge by then, filling the room and blocking their view of the Pharons.

"Into the stairwell!" Cort ordered. "Everyone. Now!"

Vasquez and Hawk did a quick, crouched run the twenty steps to the open door and went through right behind Marva, Cort, and Hank.

Hawk kicked away the debris holding the door open and let it swing closed.

"Blow the bridge," Cort ordered.

Hawk stopped for a moment and snapped something on his belt.

The next moment the entire building seemed to shake, as if they were in another earthquake.

And then it kept shaking.

Dust and debris fell around them, choking the staircase into total darkness.

"It shouldn't be doing this!" Hawk shouted over the commlink.

"Keep moving down!" Cort shouted over the roar.

Hank had no idea what was happening. He knew the shaped charges shouldn't have had so much force. More than likely the Pharons had fired back.

Or their ship was attacking.

He snapped on his light, but it couldn't penetrate the dust.

He couldn't see a thing.

He pressed himself against the shaking wall, feeling the rumbling through his back. Then he kept going, stumbling and falling downward, using the wall for balance where he could. He just hoped that Vasquez and Hawk with Cort were behind him.

Slowly the rumbling stopped around them.

But very slowly.

Hank had made it to the next landing by the time it had completely stopped.

The visibility was no greater there. Even with the light right in front of his face he could see nothing but swirling dust.

"You make it, Doc?" Cort asked over the link.

"On the next landing down," Hank said. "Can't see a damn thing."

"We're right behind you, Doc," Hawk said. "Just keep on going."

Hank wanted to know if any of the Pharons had made it across that bridge. But he stumbled a dozen more steps down through the dust before he got the words out.

"Anything following us now?" Hank asked. "Can the motion detector get a reading?"

"Nothing but us in this building," Cort said. "Some in the staircase of the one next door, moving down. From the looks of it they're going back to the last sky bridge, over to another building and then on down."

"Shit," Hawk said. "They're going to go down and flank us."

Exactly what Hank had been thinking.

"We've got to keep everyone moving, dust or no dust," Cort said. "We've got to get to the street."

And Hank understood why. If the Pharons trapped them in the stairwell, they were all as good as dead.

"No rest until we're on the ground, people," Cort said. "Pass it on to the civilians."

"Copy that," Jenkins said.

The swirling dust was thinning some as Hank stumbled to the next staircase and started down, his back against the wall. But he still couldn't even see his hand holding his rifle.

It was slow going.

And just then, he didn't have time to go slowly.

24

Time: 6:42 P.M. Pacific Time
17 hours, 11 minutes after Arrival

Stephanie ducked out of the front door of the alien skyscraper into the dim twilight. At that time of the evening very little of the Maw's light penetrated down into the street. She turned and followed Stanton, Lee, Bogle, and Edaro down the street to the east and away from that building.

At their head was Private Jenkins.

Private Waters was behind her.

It felt great to be on the ground and headed east again. Toward the ocean. The cold air on her face refreshed her and momentarily washed away some of the fear. It seemed like a lifetime since they'd left the transport and started into the heart of the city.

Maybe two lifetimes.

With any luck, they'd soon be back at the facility. But that was going to take some real luck at this point. Hank, Vasquez, Cort, Marva, and Hawk were still fifteen stories up in the building, coming down as fast as they could go.

But from the motion detector Marva carried, it was clear the four Pharon foot soldiers were going to beat them to the street and appear just one block up to the north.

In less than four minutes.

The Pharons wouldn't beat Hank and the troopers by much, but unless the aliens could be slowed down or held off, Stephanie and the others would be trapped there, their retreat cut off.

Jenkins pointed to a pile of wreckage. "Good cover. Let's hold here."

They all gathered while Waters took up a position on his stomach facing the direction from which the Pharons were coming.

Stephanie was amazed at how exhausted and dirty they all were. It looked as if they had all walked through a dust-filled hell. Actually, they were still in it. And there was still a good chance none of them was going to survive this.

Edaro quickly pulled out his computer.

"How long?" Cort asked.

"The energy is building to levels I never thought it could reach," Edaro said. "It's like a volcano at this point. It might blow at any time. Might be another hour. Depends on what triggers it to blow."

Stephanie's stomach clamped down on nothing. They were so close to getting out, yet the entire city might shift with them at any moment. And even if they lived through the shift, they would be stuck in this dead alien city with Pharons and Sand.

She would rather have the shift kill her.

"Doctors," Jenkins said, "I'm afraid I'm going to have to ask you to head down the street on your own. Waters and I have to try to cover for the rest. Once the fighting starts here, Private Cort says that if it looks like we're losing, we run full out back to the beach. There should be someone along the way to extract you."

"Understood," Bogle said. "But I plan to stay and help. Edaro, you get that information of yours to safety." He slapped his rifle. "I've been carrying this long enough. I should try firing it a few times."

"I want to stay, too," Stephanie said.

"Actually," Edaro said, "I think we're safer right here than running alone down that street."

"I agree," Stanton said, the sweat running down his face.

"I guess we're going nowhere fast," Lee said.

Jenkins nodded, his brown eyes cold, a frown on his dust-covered face. He glanced down the street, then nodded. "You've got a point about being safer here. I don't like it, but it's going to have to do. Spread out along both sides of the street and get into good cover positions. Those damn aliens will be here in less than a minute. And keep your fool heads down and let us do the fighting."

Bogle gave a mock salute and slipped behind a good hunk of debris where he had a clear shot down the road. The others also scattered, each finding rocks or piles of rubble to hide behind.

Stephanie crouched beside Jenkins. The Pug pistol was still in her pocket, but she didn't draw it. She was sure that if she ever got close enough to a Pharon to use a pistol, she'd probably already be dead.

But her reason for wanting to stay wasn't to take part in the fighting, but to help anyone who was still alive when the fighting ended. She was the medical doctor. Her job was to stay close to the battle for the sake of the wounded.

"Here they come," Jenkins said. "How soon until you're out of there, Vasquez?"

Stephanie wished she could have heard the answer, but she couldn't.

At that moment, one block up the street, one of the most horrifying sights she had ever seen came shambling out of the building, followed by two more just like it.

Pharons.

Walking dead.

Mummies of humanoids who had somehow been reanimated, covered in beautiful, engraved golden armor and equipped with packs that cycled fluids through their dried bodies.

They looked worse in person than they had on the video playback she'd seen.

Though their forms were ugly, their armor was exquisitely covered with glyphs and symbols. She could understand why those who'd studied the Cache called them Pharons. They had a kind of Egyptian look.

The armor seemed to cover their mid-bodies and shoulders, but not their arms and legs. There it was obvious their skin was wrapped, and what skin did show was an ugly, rotting gray.

Almost instantly their odor drifted over. Stephanie had smelled plenty of corpses in her time, but nothing as heavy and as sickening as this odor of death.

Around her the other doctors covered their noses, while Jenkins and Waters didn't even seem to notice.

All of the Pharons were carrying long, curved weapons of some sort. Even the weapons were decorated with intricate symbols.

She was close enough to see that the first Pharon was missing a hunk out of one side of its gray face and was minus one wrapped hand. But it was still shambling along.

The second walked with a limp because part of its leg was missing. The aliens had apparently been hurt in the fighting, but that didn't seem enough to stop them. It looked like Pharons had to be down, armor out of commission, before they stopped.

And maybe, she guessed, sometimes not even then. Anything that was dead and still walking probably could take a lot more damage than most creatures.

Jenkins and Waters both opened fire at the same time. Stephanie could see their bullets rip at the armor of the closest Pharon, but it wasn't enough to down him.

Bogle opened up with his rifle, snapping off shots against the Pharon armor, trying for more vulnerable spots.

Right behind him Lee and Edaro opened fire.

The street around her echoed with the battle, the sounds of rifle fire echoing off the buildings.

She had never been in the middle of something like this. The sound was overwhelming. Her focus became even more intense, but she managed to remind herself to breathe.

Beside her Jenkins suddenly stood and threw a grenade before ducking back down. Stephanie watched as the grenade hit its mark and exploded right under the leading Pharon, blowing it into a dozen pieces and downing one of the two following him.

The second one got right back on its feet and kept coming, bringing its strange-looking weapon up ready to fire as it walked over a leg of the one Jenkins had just blown up.

Another Pharon now emerged from the same building as his companions turned toward them.

That was all four of them. One down, three to go.

Everyone kept firing, pouring shots into the Pharons.

Stephanie couldn't believe the creatures weren't falling. She could actually see the bullets ripping into them.

"Aim at the heads!" Stanton shouted.

The leading Pharon fired its weapon. A corner of one building exploded behind them, showering them in dust and rock.

"Head shots!" Jenkins yelled. Instantly that seemed to make a difference, as three or four of them aimed at the head of the leading Pharon soldier.

For a moment Stephanie watched the Pharon stagger, then twist sideways as the bullets ripped through the cloth wrapping covering its head. Bullets smashed into its support tanks.

More pounded its head.

Fluid sprayed from the tanks.

Then, as if a small grenade had gone off just inside the Pharon's golden armor, its head exploded.

The headless soldier in his beautiful decorated armor marched on for a dozen steps before it fell over a hunk of debris and twisted there.

The smell got even worse, making her eyes water.

Rotted flesh.

Worse than any morgue she'd ever been in.

Two aliens down and two to go.

Stephanie was very glad none of them was what Stanton had called a Pharon priest. These regular Pharon soldiers were hard enough to kill.

From out of the building near them, Hank and Vasquez came running, firing at the Pharons just down the street from them as they came.

Two Pharons got off energy shots at Hank and Vasquez. One shot exploded high against the side of the building, sending a massive cloud of debris and dust over the street. The other shot caught Vasquez solidly in the back and smashed him against the wall.

Stephanie could tell even from a distance that by the time his body hit the sidewalk, it was nothing more than a burnt-out, dead husk. From the looks of it you didn't take a direct hit from a Pharon weapon and live.

Hank was knocked off his feet by the blast. He rolled and came up running, firing back at the Pharon soldiers as he went.

"Cover him!" Jenkins shouted.

All of them opened up again at the two remaining aliens.

The noise made Stephanie's head ring, but she couldn't stop watching.

Behind Hank, Marva and Cort appeared from the door of the building, moving as fast as they could as Private Hawk covered them.

"Grenades!" Cort shouted as he ran.

Alongside Stephanie, Jenkins stood and threw another grenade.

Hawk did the same from his position covering the retreat of Hank, Cort, and Marva.

Jenkins's grenade exploded against the leg of one Pharon just as he was shooting. The energy shot went wild, smashing into the building high above them as the Pharon tumbled over backward, one leg completely gone.

Hawk threw another.

Then Jenkins.

Then Hawk.

They rained grenades on the two aliens as Stephanie covered her head. Debris seemed to be raining down like a hard summer storm. The dust made it almost impossible to see very far, and the smell was so bad it had her eyes watering like she was crying.

And the sounds of the rifles firing filled the street like a thundering storm, never really pausing.

Stephanie wiped the dirt and water out of her eyes and peeked out from her cover. Through the dust she could see that both Pharons had been knocked down by the last assault. But both were trying to get up again.

"Hit them one more time!" Cort shouted. "Grenades."

Jenkins stood and threw again. Then without ducking back down, he started blasting round after round of bullets into the downed aliens.

This time the two Pharons stayed down, as the grenades and bullets sent wrapped mummy limbs flying in all directions.

"Cease fire!" Cort ordered.

Around them the street quieted.

To Stephanie the sudden quiet seemed almost louder than the fighting. It hadn't taken long, yet the fight had seemed to go on forever.

Dust filled the air, and the smell of explosions almost overpowered even the stink of the Pharons.

Stephanie couldn't believe it had stopped.

She stared at the body of Private Vasquez. His eyes were open and gone, burnt out in the intense blast of energy. He hadn't even known what hit him. That much was good.

She jumped up and ran over to check him. Private Hawk was already there. It was too late. Vasquez was dead.

The street in front of them was littered with different Pharon parts, some still moving. Stephanie desperately wanted to go check them out, to see what they might reveal. But she had no doubt that would be suicidal. There were probably a thousand ways those Pharon soldiers could still kill her.

"Let's get out of here," Cort ordered. "Standard cover formation. Jenkins, take point."

Private Marva stayed with Cort, helping him down the street toward the ocean.

Waters moved over quickly to support Cort's other side.

"Hawk, bring up the rear," Cort ordered. "Put a sonic marker on Vasquez's body."

Hawk did as he was ordered.

Stephanie turned and followed behind Dr. Lee. All of them looked as shocked as she felt. Hank looked like he'd come through hell.

"You made it," she said to him.

"This far," he said. "But we're not out of this yet."

"Right," she said. "And the closer we get to the ocean, the more vulnerable we're going to be from air attack."

"Exactly," Hank said. "But also more likely to be picked up."

"I like that thought," she said.

Stephanie stayed right with him, moving quickly down the street. Ahead of them the other four scientists moved at the same pace, staying behind Cort and the two helping him, plus Jenkins on point.

Behind them Stephanie knew the massive phase generator was building up energy to take the city on another jump to some other corner of the Maelstrom. Or maybe outside the Maelstrom. Who could say?

What she did know was that the island had already been on Earth for over seventeen hours.

She just hoped it had another few minutes left in it.

She was desperately afraid it didn't.

25

They had managed to run and walk a good twenty blocks without further attacks by the Pharons. Hank was having a very tough time catching his breath. Though Stephanie was in somewhat better shape, it wasn't by much. They were going to have to stop for a few minutes soon. But he didn't feel they had a few minutes.

Behind him he heard Stanton swear at something again. The guy had been doing that for the last ten blocks, just about every twenty steps.

Hank ignored him and focused on where he was going. The buildings around them were down to only twenty stories or so tall. He doubted that at this level the structures offered much cover from the Pharon ship.

Jenkins was still in the lead. Marva was helping Cort just ahead of them. Hank clicked on his commlink. "We need to take a rest in the next block," he managed to pant out.

"Understood," Cort's voice came back in his ear. "First cover, take it."

Then Cort did something on the commlink Hank hadn't heard before.

"Private Cort to mainland, come in please?"

Nothing. Or at least nothing that Hank could hear.

"This is the Union squad on the alien island," Cort said. "Sergeant Malone is dead. We need retrieval quickly. Island is about to phase to a new location. Evacuate all other personnel."

Again nothing but silence in Hank's ear. Hank had no idea if Cort was getting an answer or just sending the signal in hopes that someone would pick it up. More than likely, the buildings were still too tall for a comm signal to get out.

Cort motioned for Hank and the rest to stop near a pile of fallen rubble near an open hole in a building. Hank managed to lower his aching body onto a broken chunk of wall. He doubted there was a place on him that wasn't banged or bruised. But at least he was still alive, which was more than Vasquez could say.

Or Malone. Or the others who'd died.

Hank wasn't letting himself think about Vasquez getting hit, knowing it could just as easily have been him. He took a few, long breaths and tried to clear some of the dust from his lungs.

Vasquez had been a good soldier. So had Sergeant Malone and the others. Hank just hoped the data they'd gathered from the massive phase generator was worth good soldiers' lives. Now they just had to get it off the damn island.

Marva and Waters lowered Private Cort down to a sitting position near Hank. Neither of them seemed to be panting at all. Or even the slightest bit winded.

Chop Edaro dropped down beside Hank, working to catch his breath. Then he got out his sensing equipment and studied it. Hank really didn't want to know what he was seeing. There was no doubt they didn't have much time left. That was all he needed to know.

He turned to Cort. "Anything on the motion sensor?"

"Trust me," Cort said, "if there was, we'd still be moving."

"Maybe they're still in the energy building," Bogle said. "Could we get so lucky?"

"I doubt it," Cort said simply.

"Get through to the mainland?" Hank asked, changing the subject.

"Don't know," Cort said. "These buildings are still pretty high, but they might have heard me."

"I hope so," Hank said. "I'm not really up for swimming back."

"They're giving support," Cort said. "Even though we can't contact them. My guess is they've been dealing with that Pharon ship. They may not be able to contact us, but they know where we are."

"You're sure of that?" Hank asked, hoping Cort was.

"Very sure," Cort said. "The Union takes very good care of its troops. Especially on a mission like this."

"Can they get a craft into this narrow road to extract us?" Hank asked.

"Not easily, but it could be done."

"I just hope they do it before the island jumps," Bogle said. "I hate swimming."

"If this island leaves suddenly, and we're offshore," Lee said, "we won't be swimming, we'll be surfing the tsunami."

"Another one?" Stephanie asked.

Hank glanced at her. She seemed to be breathing normally again.

"Of course," Lee said. "This island phases away and the water will come rushing into the void. Massive waves."

"Just great," Stephanie said.

"People," Edaro said, "we don't get off here soon, it's not going to matter."

"What's happening?" Hank asked.

"I'm getting energy spikes," he said. "My guess is the Pharons have killed a few of the Sand. These energy spikes are just like what happened when we killed those last ones."

"You'd think the Pharons would have that figured out," Hank said.

"Unless they're not studying the generator so much, but instead are trying to figure out how the Sand are re-formed," Lee said.

"That's all we need," Bogle said. "Nasty-smelling aliens who not only can come back from the dead, but re-form."

That thought made Hank's stomach twist.

"Who knows," Edaro said. "But too many more of those and we're going for a ride with this place."

"Let's move it," Cort said, motioning for Marva to help him again. "We still have a way to go to get to the water."

Within a few seconds they were again moving down the street, weaving in and out of debris, making the best time they could toward the ocean.

Two blocks later Cort tried to raise the mainland again.

This time he got a response.

"Take cover," base control said. "Transport on the way, but enemy ship approaching. We'll continue to try to hold it off, but it's hard to say if we can."

Those were not the words Hank had been hoping to hear. Not at all.

The Pharon ship had found them.

And it seemed the Union aircraft had been engaging it for some time without success.

"Take cover!" Cort ordered. "Scatter, everyone!"

Hank saw it coming from the east, the very direction toward which they were running. It was a small ship, more than likely a fighter of some sort. It was gold, with elaborate markings and symbols on the sides. It seemed to gleam in the Maw's light.

As he watched, stunned, a Union jet flashed in from the left, firing a missile at the enemy ship.

It was a direct hit, but the Pharon craft just kept on

coming. It was as if nothing had happened, and the craft seemed to drain off the energy from the missile explosion like water off a roof.

The ship was shaped like a sleek wedge, with a pointed-down nose like a bird's beak. As it got closer they could see that its body was gold and chrome, with black trim. And like the Pharon armor, it had that same Egyptian feel.

"From the east!" Hank shouted, just as the thunder from the missile explosion echoed over the city.

The ship was suddenly almost on top of them.

Then it fired point-blank at Waters and Jenkins. Neither man had a chance as the energy beam exploded in the street in front of them, sending their bodies flying like rag dolls.

The ship's momentum carried it over them as they all scrambled for cover.

Marva and Cort went for a large pile of rubble, while the rest were diving for anything even vaguely resembling protection.

The ship turned back.

Hank shoved Stephanie down a side street, then through a hole into a building. Edaro, Bogle, and Lee all piled in after them, scrambling away from the opening like rats from a sinking ship.

Outside Hank could hear the troopers blasting back at the ship, though it must have been like firing at a fighter jet with a peashooter.

"Mainland, come in," Cort shouted into the commlink. "We're under attack from a Pharon small fighter. Need help and removal immediately from the island."

"Copy that," the voice said in Hank's ear.

The sounds of two Union fighters flashing overhead shook the building.

Then massive explosions shook the ground and filled the room with light dust. Hank hoped it was the sound of the Union planes downing the Pharon ship. But there were only so many planes that could attack one small ship at once. And the Pharon craft seemed to have a shielding that drained away any kind of energy.

Hank glanced around. The building where they'd taken shelter was, at most, eight stories tall. A few good direct hits by the Pharon energy weapons and they would be buried in there.

Outside, another massive explosion rocked the ground.

More return fire from Cort.

More roaring of jets overhead.

Outside the world was going crazy.

Hank scrambled back to the hole in the wall and watched.

Above the street the sky was filled with Union jets, at least a dozen, making run after run at the Pharon ship. Missiles were slamming into it, one after another.

Another large explosion sounded in the street as the Pharon ship fired down near Cort's location. The Pharon ship was pounding them hard.

Still another massive explosion followed, this time against the side of the building they were in. It knocked him backward and head over heels.

Stephanie was at his side as he scrambled back to his feet.

"You all right?"

Dust swirled around them as he nodded. "Fine," he said.

More explosions on the street outside. Hank had no idea how they were even managing to withstand such an attack.

Then, suddenly, as if the world had completely shattered, the ground shook and bits of the roof above them broke loose. The explosion was so loud that Hank wondered if he would ever hear again.

Then the fighting seemed to stop.

The echoes of the explosions faded off into the distance as they all stood in the dark, listening.

Hank was almost afraid to breathe.

Then suddenly a voice spoke in the commlink in his ear.

"Island team," the base said. "Alien ship down. Stand by for extraction."

"The Pharon ship is down," Hank said to the others. "They're coming to get us out."

The scientists around him looked stunned.

Then he realized that Cort hadn't replied.

"Cort, did you hear that?"

There was no answer from the commlink.

"Cort?"

Dead silence.

26

To Stephanie, the attack of the Pharon ship was almost the final straw. She didn't know how much more she could take or how much longer she could even keep going. She'd never been in the military, never been around war before. She was a doctor, trained to save lives.

The explosions shook the building as she huddled against the base of the ramp, light off, trying to not move. Edaro had gone to the other side of the ramp while Bogle had gone to the far corner and Lee to the near corner.

None of them dared say a word.

None of them dared to move except Hank. He had poked his head back out the hole and was watching the battle.

Above her a massive explosion pounded the building, raining debris down on her like hail.

Hank had been knocked back, and she'd gone to help him. But he was all right.

Then the world exploded, the ground shook, and she was bounced hard off the floor. Her shin took a nasty hit when she came back down.

Something outside had blown up.

Something big.

Could Private Cort have managed to destroy the Pharon ship? She didn't know how he could have.

Dust filled the air, and she coughed, but didn't move.

They didn't dare move.

Suddenly through the dust she heard Hank saying the alien ship was destroyed and it was clear.

Then a moment later, he said, "Follow me. Quickly. I think something's happened to Cort."

Hank ducked back out of the hole in the side of the building. Stephanie scrambled to her feet and followed him.

The street outside didn't look anything like it had just a few minutes ago. Debris filled it, and two of the nearby buildings had completely collapsed.

The center of one building was on fire, and after a moment Stephanie could tell what was burning. The Pharon ship.

A half dozen Union planes were in the sky overhead. Then suddenly two Union war planes streaked past very low, their thunder ripping through the city.

Now she understood what had happened. The Union had downed the Pharon ship. She almost wanted to cheer those two planes like a schoolkid at an air show.

Hank scrambled back into the street, clearly looking for any survivors from the squad as Stephanie followed, shocked by what she was seeing. A part of one man's leg had been ripped off and smashed against a wall. She didn't even want to think whose leg it might have been.

Where Cort and Marva had been firing from was now a small crater in the street. There was no chance they could have survived that kind of direct hit. Dr. Stanton was also dead, curled in a ball and smashed by falling wall debris.

Stephanie could barely take it in. She felt numb all over. Not one of the Special Ops troop had survived. They had all given their lives for this mission.

And to keep her and the rest of them alive.

How could that be possible?

How could anything be worth that?

From over the top of one building to Stephanie's right a Union Hydra transport appeared, lowering itself carefully between the two buildings, barely fitting in the street. Under normal circumstances, the pilot would have never tried that. But this clearly wasn't a normal circumstance.

Not even close.

The ramp on the craft opened, and Hank shouted from beside her, "Everyone aboard."

Stephanie climbed and stumbled over some rocks and piles of debris to follow Dr. Edaro into the transport. A soldier in there pushed her gently into a seat and helped her get her belts buckled. She let him, sitting there numb and feeling drained.

She would have thought she would be excited to be rescued. But at the moment she wasn't feeling anything.

It was safer that way.

Hank, Lee, and Bogle all followed on board and got strapped into seats.

"That everyone?" asked the trooper who was strapping them in.

Hank nodded slowly.

"Shit," the soldier said.

Then the transport door closed with a final-sounding thump.

They had been rescued.

Five civilians.

No soldiers.

The transport lifted, then turned and headed east, picking up speed.

Stephanie looked over at Hank, who was sitting there staring down at the floor.

Dr. Edaro was shaking, the golf ball gripped tightly in his hand.

Dr. Bogle was just smiling. A smile of complete shock that they had actually survived.

Lee looked as shocked as she felt, his normal smile completely gone.

She couldn't even muster a smile for the moment of rescue.

Or a tear for all the good soldiers who had died.

So she just sat there, staring at the floor, trying not to think or remember anything. But Private Cort's face kept coming to her mind.

And behind his face was Sergeant Malone's. They had done their duty. They had died for it. Stephanie just hoped they had found peace.

"It's going," Edaro said, staring down at the open sensor on his lap.

Hank and Bogle both turned to look out the small window on the transport.

Stephanie didn't want to watch. She didn't want to even remember.

"Gone," Edaro said. "Eighteen hours and twenty-one minutes after it arrived." He was studying his instruments as he spoke. "Took that damn Pharon high priest with it. From the looks of the energy spikes, they must have killed a hundred Sand."

"Serves the bastards right," Hank said softly.

Stephanie said nothing. They had gotten out just in time. Only moments to spare. She couldn't believe it.

Knowing the island was gone helped a little. She could feel the weight lifting. The relief edging in sideways.

She glanced out the window over Hank's shoulder. The ocean was empty. But there was a bright, odd light seeming to fill the air everywhere.

The waters were swirling below, and lightning was flashing across the surface, just as it had done when the island arrived. Inside the transport, she couldn't hear the thunder, and she was glad for that.

Then, as she watched, the lightning stopped.

And the weather below them seemed to clear.

The odd light faded and was gone, just as it had done when the island arrived.

But this time the ocean remained empty.

As it should be.

27

Time: 11:37 P.M. Pacific Time
6 Days, 3 hours, 45 minutes after Departure

H ank moved along the beach, the sand filling
his shoes like it had a little less than a week
before, the night the island appeared. Only
this time, when Stephanie suggested a walk, he'd been
smart enough to remember to wear his heavy coat.
And his gloves and stocking cap.

And tonight he needed them all. A cold wind blew
off the ocean, cutting through even his heavy jacket
like a knife. The surf pounded on the shore, filling the
air with the smell of salt and the rumbling sound of
waves crashing against a beach.

He'd been surprised, and pleased, when Stephanie

had suggested the walk, They'd both spent the last six days since being rescued sleeping and being debriefed by everyone who had even thought of a question. And, knowing the military, the questions were a long way from being over.

During the last six days Stephanie had been distant and aloof. Hank imagined he hadn't been much better. What had happened out on the island wasn't going to be easy for either of them to put to rest.

He knew for a fact he would never forget it.

But the mission, even with such heavy casualties, had been a success. The information Edaro and Lee brought back would keep a small army of Union researchers busy for decades, Hank figured. And who knew how much benefit the Union, and all of humanity, would get from what they recovered from the alien city. Edaro even thought that the massive phase generator might be a way out of the Maelstrom.

Hank figured anything was possible from the island-city.

Still, as far as he was concerned, that was small consolation for all the lives lost. Sergeant Malone, Private Cort, and the others. Hank felt proud to have known them.

And damn lucky they'd been there to protect him and Stephanie.

He was sure Stanton's expertise and fine scientific mind would be very much missed by the Union. It wasn't until after they got back that Hank learned that Stanton had played a key role in the research on the Cache.

Stephanie reached over and took his arm, pulling

him close as they walked. "Beautiful night, isn't it?" she asked.

He glanced at the starless blackness of the Maelstrom night. "Sure is."

"You just saying that?" Stephanie asked.

"Nope," he said. "After this last week, walking like this on a beach with you will always mean I'm having a beautiful night."

She laughed softly, but said nothing.

They walked slowly for a short time, then stopped and faced out over the blackness of the night sea.

"It's never coming back, is it?" Stephanie asked.

"I doubt it," Hank said. And he did. He was telling her the truth. The island wasn't coming back, at least not to this spot. There were just too many places out there in the Maelstrom for it to appear.

But he wasn't so sure about the Pharons. Since they'd returned to the mainland, he'd learned a lot more of what the Union had found in the Cache. His bet was that humanity hadn't seen the last of those walking dead by a long shot. And they'd been damn lucky to escape them. Again, the fighting ability, bravery, and intelligence of Malone and Cort had been the key.

"So how do I get that island to leave my head?" Stephanie asked.

"You don't," he said. "You just leave it there, work around it, live with it."

"Like a tumor that isn't growing, but isn't shrinking," Stephanie said.

"Yeah," he said, not liking the image at all. But he supposed it was accurate.

They stared in silence out over the black ocean as the waves pounded the beach in front of them. Then, finally, Stephanie took his arm and turned them both back toward home. Then, just before they took the path toward the facility, Stephanie stopped and turned to look out over the ocean one more time.

"It's gone."

Hank knew it wasn't a question. It was just a statement.

"It's gone," he said.

But the memory, the data gleaned from its short presence on Earth, would live on in the Union. As would the memory of those who'd died to bring that treasure home.

It was the way of war. Hank knew that. He didn't like it, but he knew it.

"Let's go," Stephanie said, turning away from the ocean. "I've got some work to do tomorrow."

"So do I."

He gave her a hug as they walked up the sandy trail, the ocean, the cold wind, and the memory of the island at their backs.

DEAN WESLEY SMITH is the best-selling author of over forty novels and hundreds of short stories. He's written novels in the gaming world including *Shadow Warrior* and *Unreal*. He did the original script for the live action game *Star Trek: Klingon*, as well as the novelization. His most recent novel is *Men in Black: The Green Saliva Blues*. Dean has also been a publisher and editor and is currently editing a Star Trek anthology called *Strange New Worlds* for Pocket Books. His work has been nominated for every major award in science fiction and fantasy, including the World Fantasy Award, Nebula Award, Stoker Award, Locus Award, and four times for the Hugo Award. He has won both the Locus Award and a World Fantasy Award.

VISIT WARNER ASPECT ONLINE!

THE WARNER ASPECT HOMEPAGE
You'll find us at: www.twbookmark.com then by clicking on Science Fiction and Fantasy.

NEW AND UPCOMING TITLES
Each month we feature our new titles and reader favorites.

AUTHOR INFO
Author bios, bibliographies and links to personal websites.

CONTESTS AND OTHER FUN STUFF
Advance galley giveaways, autographed copies, and more.

THE ASPECT BUZZ
What's new, hot and upcoming from Warner Aspect: awards news, best-sellers, movie tie-in information . . .